Julian Edmund Tenison-Woods

**Fish and Fisheries of New South Wales**

Julian Edmund Tenison-Woods

**Fish and Fisheries of New South Wales**

ISBN/EAN: 9783337329624

Printed in Europe, USA, Canada, Australia, Japan

Cover: Foto ©berggeist007 / pixelio.de

More available books at **www.hansebooks.com**

# FISH AND FISHERIES

OF

# NEW SOUTH WALES.

BY THE

REV. J. E. TENISON-WOODS, F.L.S., F.G.S., &c., &c.,

VICE-PRESIDENT OF THE LINNEAN SOCIETY, NEW SOUTH WALES.

AUTHOR OF "A HISTORY OF THE DISCOVERY AND EXPLORATION OF AUSTRALIA,"
"GEOLOGICAL OBSERVATIONS IN SOUTH AUSTRALIA,"
"NORTH AUSTRALIA,"
"NATURAL HISTORY OF NEW SOUTH WALES,"

SYDNEY: THOMAS RICHARDS, GOVERNMENT PRINTER.

1882.

2a 56-82

# Dedication.

TO

## FREDERICK M'COY, F.R.S., F.G.S.,
MURCHISONIAN AND CLARKEAN MEDALLIST,
PROFESSOR OF NATURAL HISTORY AT THE MELBOURNE UNIVERSITY, &c., &c.,

### THIS WORK IS INSCRIBED

AS AN

ACKNOWLEDGMENT OF HIS EARLY

LABOURS ON BEHALF OF PALÆONTOLOGICAL SCIENCE IN EUROPE,

AND HIS

GREAT SERVICES TO AUSTRALIAN NATURAL SCIENCE

DURING THE LAST EIGHTEEN YEARS,

BY HIS OBLIGED FRIEND

THE AUTHOR.

# PREFACE.

THIS work has been undertaken with a twofold object. The first is to give a popular account of our useful Fishes to the colonists, together with all that relates to pisciculture and acclimatization, with a view to promote a development of our fish resources. To make it more useful in its educational character, a simple explanation of the science of the subject is prefixed.

The second object is to give an account of our Fish and Fisheries, such as will be required for the great Fish Exhibition of 1883. With this view, all that relates to our laws, markets, and our fisheries generally is given.

Finally, the work is meant to be simple and practical, so that not a word in it may be above the comprehension of those not specially trained to scientific phraseology. Though in places necessarily a little technical, it is hoped that on the whole it will be found readable, and sufficient to give to inquirers all the knowledge they require about the Fisheries of New South Wales.

I take this opportunity of thanking the Hon. W. Macleay, F.L.S., and Alex. Oliver, Esq., for valuable assistance in revising these pages, and also Prof. M'Coy, F.R.S., for permission to use some of the plates of his work on the Zoology of Victoria.

# CONTENTS.

|  |  | PAGE. |
|---|---|---|
| DEDICATION | | i |
| PREFACE | | iii |
| LIST OF PLATES | | vii |
| GLOSSARY | | ix |
| CHAPTER I. | INTRODUCTION | 1 |
| II. | THE FISH FAUNA OF NEW SOUTH WALES | 10 |
| III. | OUR MARINE FOOD FISHES | 30 |
| IV. | SHARKS | 92 |
| V. | RAYS | 99 |
| VI. | OUR FRESH-WATER FOOD FISHES | 102 |
| VII. | OYSTER FISHERIES | 110 |
| VIII. | OTHER MOLLUSCA | 122 |
| IX. | CRUSTACEA | 125 |
| X. | FISHING-GROUNDS OF NEW SOUTH WALES | 128 |
| XI. | THE FISH MARKET | 139 |
| XII. | THE DEVELOPMENT OF OUR FISHERIES | 143 |
| XIII. | ACCLIMATIZATION AND PISCICULTURE | 150 |
| XIV. | FISHERY LAWS AND REGULATIONS | 160 |
| XV. | INDEX OF LOCAL NAMES | 182 |
| XVI. | WORKS RELATING TO FISH AND FISHERIES | 194 |
| INDEX | | 205 |

# LIST OF PLATES.

| PLATE. | | PAGE. |
|---|---|---|
| Frontispiece. | "Old Man" Schnapper. | |
| I. | The Perch. *Lates colonorum* ... ... described at | 31 |
| II. | The Old Wife. *Enoplosus armatus* ... ... ,, | 32 |
| III. | Longfin. *Anthias longimanus* ... ... ,, | 33 |
| IV. | Wirrah. *Plectropoma ocellatum* ... ... ,, | 34 |
| V. | Australian Salmon. *Arripis salar* ... ,, | 35 |
| VI. | Sweep. *Scorpis æquipennis* ... ... ,, | 37 |
| VII. | Black-fish. *Girella tricuspidata* ... ... ,, | 39 |
| VIII. | Schnapper. *Pagrus unicolor*... ... ... ,, | 39 |
| IX. | Tarwhine. *Chrysophrys australis* ... ... ,, | 42 |
| X. | Morwong. *Chilodactylus macropterus* ... ,, | 46 |
| XI. | Red Morwong. *Chilodactylus fuscus* ... ,, | 46 |
| XII. | Red Rock Cod. *Scorpæna cardinalis* ... ,, | 47 |
| XIII. | Banded Morwong. *Chilodactylus vittatus* ... ,, | 47 |
| XIV. | Gurnard Perch. *Sebastes percoides* ... ... ,, | 48 |
| XV. | Nannygai. *Beryx affinis* ... ... ... ,, | 51 |
| XVI. | Jew-fish. *Sciæna antarctica* ... ... ,, | 53 |
| XVII. | Teraglin. *Otolithus otolodus* ... ... ,, | 54 |
| XVIII. | Yellow-tail. *Trachurus declivis* ... ,, | 58 |
| XIX. | King-fish. *Seriola lalandii* ... ... ,, | 59 |
| XX. | Tailor. *Temnodon saltator* ... ... ,, | 60 |
| XXI. | John Dory. *Zeus australis* ... ... ,, | 61 |
| XXII. | Horse Mackerel. *Trachurus trachurus* ... ,, | 58 |
| XXIII. | Whiting and Beardie. *Sillago maculata* and *Lotella marginata* ... ... ... ,, | 65 |
| XXIV. | Northern Whiting. *Sillago ciliata* ... ,, | 65 |
| XXV. | Flathead. *Platycephalus fuscus* ... ... ,, | 67 |
| XXVI. | Gurnard and Sergeant Baker. *Trigla kumu* and *Aulopus purpurissatus* ... ... ,, | 68 |
| XXVII. | Flying Gurnet. *Trigla polyommata*... ... ,, | 68 |
| XXVIII. | Pike. *Sphyræna obtusata* ... ... ,, | 69 |
| XXIX. | Sea Pike. *Lanioperca mordax* ... ... ,, | 69 |
| XXX. | Sea Mullet. *Mugil grandis* ... ... ,, | 61 |
| XXXI. | Blue Groper. *Cossyphus gouldii* ... ... ,, | 74 |
| XXXII. | Pig-fish. *Cossyphus unimaculatus* ... ,, | 75 |
| XXXIII. | Flounder. *Pseudorhombus russellii* ... ,, | 76 |
| XXXIV. | Sole. *Synaptura nigra*... ... ... ,, | 77 |

## LIST OF PLATES.

| PLATE. | | PAGE. |
|---|---|---|
| XXXV. | Sergeant Baker. *Aulopus purpurissatus* ... described at | 82 |
| XXXVI. | Long Tom. *Belone ferox* ... ... ... ,, | 83 |
| XXXVII. | Garfish and River Garfish. *Hemirhamphus intermedius* and *H. regularis* ... ... ... ,, | 84 |
| XXXVIII. | River Garfish. *Hemirhamphus regularis* ... ,, | 84 |
| XXXIX. | The Eel. *Anguilla australis* ... ... ... ,, | 88 |
| XL. | Leather-jacket. *Monacanthus ayraudi* ... ,, | 89 |
| XLI. | Murray Cod. *Oligorus macquariensis* ... ... ,, | 102 |
| XLII. | Sea Crab. *Neptunus pelagicus* ... ... ... ,, | 125 |
| XLIII. | Cray-fish. *Palinurus hugelii* ... ... ... ,, | 125 |
| XLIV. | Freshwater Cray-fish. *Astacopsis serratus* ... ,, | 126 |
| XLV. | Bat-fish. *Psettus argenteus* ... ... ... ,, | 61 |
| A wood engraving of heads of *Salmo*, with operculum and teeth | | 9 |

# GLOSSARY.

**Acuminate.**—Tapering to a point.
**Adipose.**—Fatty matter throughout the tissue.
**Armature.**—A prickle or bony point.
**Axil.**—Literally the armpit; the inner angle at the base of a fin or spine.
**Basibranchials.**—Small bones uniting the branchial arches below.
**Branchial.**—Relating to the gill.
**Branchiæ.**—Gills.
**Branchial arches.**—Plates which bear the gills. In the Perch they are five, four bearing gills; the fifth dwarfed, and bearing teeth called lower pharyngeal bone. The arches are divided into movable portions dilated and confluent above, beset with fine teeth, and called the upper pharyngeal bones.
**Branchiostegals.**—The margin of the gill-cover has a skinny fringe to close it more effectually; this fringe is supported by one or many bony rays called branchiostegals.
**Bifurcated.**—Forked.
**Bumbora.**—A sunken rock or reef covered at low-water.
**Canines.**—Larger projecting teeth like dog teeth.
**Ceratobranchials.**—The middle segments of the branchial arches.
**Cœcal.**—Blind, or closed sac, not perforate.
**Confluent.**—When separate parts become united so as to lose their distinctness they are said to be confluent.
**Cran.**—A barrel of herrings.
**Deciduous.**—Easily falling away.
**Denticulate.**—Finely toothed.
**Dun.**—Broken dried cod-fish.
**Emarginate.**—Notched or cut out at the edge.
**Epibranchials.**—The upper segment of the branchial arch.
**Ethmoid.**—A cartilage, thickest above the vomer, which extends as a narrow stripe along the bony partition between the eyes. The nerves of smell (olfactory) run along and through the Ethmoid.
**Falcate.**—Crescent-shaped or bent like a reaping-hook.
**Filament.**—A thread-like membrane.
**Finlets.**—Small fins, such as those along the tail of the mackerel.
**Follicles.**—Small leaf-like bags or cavities.
**Gibbous.**—Swollen or humped at some part of the surface.
**Grilse.**—Young salmon that have never spawned.
**Hill or To Hill.**—Sand-heaps formed by Salmon for the ova. Hence the expression "to hill," for spawning time. The sand-heap is also called a ridd.
**Humeral.**—Belonging to the shoulder.
**Hyaline.**—An extremely clear, transparent membrane.
**Hyoid.**—The arch which encloses the branchial apparatus.
**Hypobranchial.**—The lowest segment of the branchial arch absent from the fourth arch.
**Imbricate.**—Laid over each other like tiles.
**Incisors.**—Front cutting-teeth in fishes.
**Kelt.**—Salmon exhausted, and often covered with parasites after spawning-time.

# GLOSSARY.

**Klipfish.**—Dried cod.
**Lamellæ.**—Thin plates.
**Laminate.**—Divided into thin plates.
**Lateral.**—Pertaining to the side; refers especially to the line along the sides of some fishes called the lateral line.
**Littoral.**—Belonging to the shore.
**Maise or Maese.**—A measure of 500 herrings.
**Maxillary.**—The front margin of the upper jaw formed by the intermaxillary or premaxillary bone which in most fishes bears teeth. It is spread into a flat, triangular projection, on which leans the second bone of the upper jaw—the maxillary.
**Mesially.**—Mesial; a supposed divisional, perpendicular, longitudinal line, dividing the fish into two halves.
**Muciferous.**—Mucous system, having special reference to the lateral line, the scales of which are perforated and exude an oily lubricating fluid.
**Nancy.**—A trade term for 40 lobsters.
**Occiput.**—The hinder part of the head or skull.
**Offal.**—Trade term for Haddock, Plaice, and Whiting.
**Olfactory.**—Relating to the organs of smell.
**Operculum.**—Gill-cover.
**Orbit.**—The eye socket.
**Osseous.**—Bony tissue.
**Ossicles.**—Small bones.
**Palatine.**—Relating to the palate bones. The palate properly consists of three bones—1. The Enteropterygoid, an oblong thin bone attached to the inner border of the palatine and pterygoid, and increasing the surface of the bony roof of the mouth. 2. The Pterygoid or transverse bone joined by suture to the. 3. Palatine, which is generally toothed and joined to the vomer.
**Parr.**—The first stage of young salmon.
**Partan.**—A Scotch provincial name for the common edible British crab, *Cancer pagurus*.
**Pelagic.**—Inhabiting the open ocean.
**Pharynx.**—The gullet.
**Pharyngeal.**—See "Branchial arches."
**Pink.**—Another name for smolt.
**Plicate.**—In folds or ridges, or plaits.
**Preoperculum.**—A bone with a free margin on the operculum.
**Prime.**—Sole, Turbot, Brill, and Cod, trade term.
**Pseudo-branchiæ.**—False gills. In the young stage of many fishes they breathe by a gill which becomes subsequently disused. It remains in the form of a fifth or false gill in front of the others, but not in all fishes.
**Putt or Putcher.**—Wicker-basket nets.
**Rudd, Ridd, or Redd.**—Male salmon, and the nest.
**Scutes.**—Small bony plates replacing scales.
**Segment.**—A division or joint.
**Septum.**—A partition.
**Serrated.**—Toothed like a saw.
**Setiform.**—Shaped like a bristle.
**Shotten.**—A female fish from which the roe is just discharged.
**Skegger.**—Young salmon.
**Slat.** *See* **Kelt.**
**Smolt.**—Young salmon in the very young state after the "Parr" stage, when they assume scales and lose the transverse markings.
**Spinous.**—Thorny or with bony spines.

**Strikes.**—Salmon that have spawned.
**Suture.**—Seam or division.
**Truncate.**—Ending abruptly, as if cut off.
**Trunk.**—The body of the fish, including sides, belly, and back.
**Villi.**—Minute velvety projections.
**Villiform.**—Very fine or minute conical teeth, arranged in a band.
**Vitreous.**—Glassy, clear.
**Vittate.**—Striped.
**Vomer.**—The thin bone dividing the nostrils, the base of which is often armed with teeth.
**Xiphoid.**—Shaped like a sword.
**Zonate.**—With zones or bands of colour.

Any terms not found here are explained in the context where they are used.

# Fish and Fisheries of New South Wales.

## CHAPTER I.
### Introduction.

THE fisheries of this Colony have recently attracted much attention; they have become an industry increasing each year in extent and value, but it is only within a year or so that anything definite has been known about them. One or two private individuals have interested themselves, but, until the Royal Commission on the Fisheries of New South Wales in 1880, any reliable and systematic information was not within the reach of the public. Now that the defect has been remedied it is proposed to give a popular account of our fish and fisheries, for the benefit of the colonists generally or those interested in the subject elsewhere. As far as the facts are concerned nothing would be better than the Blue Book published by the Commission, containing the report, the minutes of evidence, and the valuable appendices. The report especially is admirable in every way, and abundant use will be made of it in this essay; but something more is requisite to make the scientific side of the subject a little more accessible. We have popular manuals of botany and geology, in which the technical terms in use are carefully explained and made familiar. The terms in both these sciences are numerous and perplexing; it is not so with the science of fishes, or as it is called Ichthyology. The technicalities are few and easily understood, and when they are so understood the scientific manuals on the subject will be open to all. By this means every fisherman of ordinary education will be able to get the scientific name of any species he may find. This would seem but a small gain, but when it is known that under this name can be found all that science has recorded about the fish, its uses, habits, structure, anatomy, and its place in nature, the gain of knowledge is great. Fishermen no doubt have their own names, and in a rough way apply their experience; but this has no acknowledged record or definition, and the local name of a species here may be applied to a totally different fish elsewhere. All the facts of experience may easily become misapplied through a name which thus becomes a source of confusion instead of knowledge; but with a recognized name it is hardly possible that an intelligent fisherman will not be largely benefited by the scientific researches of others. It is not now as it was a few years ago, when such a thing as a scientific catalogue of fishes was not to be found, and even the incomplete works on the subject were enormously expensive and quite out of the reach of persons of ordinary means. Since Dr. Günther published his great Catalogue of Fishes the subject has been placed within the reach of all. There are many copies of this

A

work in the Colony, both in public institutions and in private libraries. Besides this, the Hon. W. Macleay has published a complete Catalogue of Australian Fishes, and thus all that science has done can be easily and thoroughly known. This catalogue adds more than 600 to the number of known Australian species, which is now about 1,150.

It is not intended in this essay to give a detailed description of all our fishes, but merely figures and descriptions of the more common ones, or those useful kinds which employ the industry of our fishermen and fill our markets. Such information will be added as to the habits and mode of capture as will be useful either to the professional fisherman or the sportsman. All that is known of our fishing-grounds will be given too. This, with any knowledge that we have of the economical uses of our fishes, with our fishing laws, and all the statistics available, will it is hoped make this essay a complete handbook of our New South Wales Fisheries.

*Literature.*—First of all it may be useful to state what has been hitherto done in this direction. In 1870 Mr. Alex. Oliver published, in the "Industrial Progress of the Colony," a paper on the Fisheries of New South Wales. In 1874 the late Mr. Edward Smith Hill wrote a series of fourteen articles for the *Sydney Mail*, entitled "Fishes of and Fishing in New South Wales." These are most interesting and valuable, and will be often referred to in this essay. In 1877 there was an Oyster Commission appointed for N. S. Wales, which published a report. Besides these works and the Report of 1880 already referred to, there are the special writings on the subject of Australian Fishes by Count F. de Castelnau, published either in the Journal of the Linnean Society of New South Wales or in the Proceedings of the Zoological Society of Victoria for 1872 and 1873. There were also two scarcely less valuable or extensive essays published in the official reports of the Victorian Intercolonial Exhibitions for 1873 and 1876. The work of the Hon. W. Macleay has been already referred to; but it may be mentioned that, though appearing first in the Proceedings of the Linnean Society of New South Wales, it was also published separately in two volumes for private circulation. Finally, there is an edition of the new Fisheries Act by A. Oliver. No other original writers need be mentioned, because the works of Cuvier and Valenciennes, Sir John Richardson, and Dr. Günther have no special reference to Australia, and the Australian fishes described in them occur casually. It is necessary to state, however, that no complete knowledge of any species can be obtained without reference to the great Catalogue of Fishes by Dr. A. C. L. G. Günther, F.R.S., published in eight volumes by the Trustees of the British Museum, or the same author's work on the Study of Fishes, which is an augmented reprint of his article on Ichthyology in the ninth edition of the Encyclopedia Britannica.

*Definition.*—It will be necessary first of all to state what is a fish. Some readers will be surprised, if not amused, at such an idea. As if everybody did not know what a fish is. Yet a fish has peculiarities which very few know, therefore a valuable increase to our knowledge will be gained by the definition. A fish is a vertebrate animal living in water, and breathing water by means of gills. It has cold red blood, but circulated by means of a heart with only two chambers, a ventricle and an anti-chamber or auricle or bulb. The limbs when present are fins, either in pairs along the sides or in a single line above and below. The skin is either (1) naked, (2) covered with scales, (3) with bony plates, (4) or bony armour. With few exceptions fishes propagate their species by means of eggs, and are hence called oviparous.

Before proceeding to an explanation of the terms necessary to understand scientific books about fishes, it may be well to refer to a plan for their classification which is useful to remember. M. Louis Agassiz proposed to arrange all fishes into four great classes according to the structure of the scales or bony covering. Thus there were 1. PLACOIDS: Without proper scales, but instead plates of enamel either large or reduced to mere points: Sharks, Rays, &c. 2. GANOIDS:—Scales angular, bony below, enamelled above. There are few living species of these, but there were many in former times. Our Australian *Ceratodus* is one, and the Sturgeon is another. 3. CTENOIDS:—Scales rough with comb-like teeth at their free margins, such as in Perch, Soles. 4. CYCLOIDS:—Scales smooth without teeth at the hind margin: Salmon, Mullet, Herring, Cod, &c. This classification is not adopted now, but it is found useful in many ways in determining the zoological character of a fish, and is moreover easily borne in mind.

A more elaborate system of classification, and one generally adopted, is that of Günther, which divides the class Fishes into four sub-classes. 1. PALÆICHTHYES: Sharks, Ganoids, and Rays. 2. TELEOSTEI, which includes the majority of fishes. 3. CYCLOSTOMATA: Lampreys and a few other rare genera. 4. LEPTOCARDII. These divisions are more natural than any others, but they are founded on minute details of anatomy which require special knowledge to determine. Thus, PALÆICHTHYES are fishes with a contractile auricle to the heart; intestine with a spiral valve; optic nerves not crossing. 2. TELEOSTEI:—Fishes with an auricle which does not contract, intestine not spiral, optic nerves not crossing. Skeleton composed of bone, with vertebræ separated completely. 3. CYCLOSTOMA:—No auricle to the heart, intestine simple. Skeleton of cartilage instead of bone. Only one nasal aperture. No jaws, but the mouth surrounded by a circular lip. 4. LEPTOCARDII:—Heart reduced to mere pulsating sinuses in the great artery. Intestine simple. Skeleton partly cartilaginous and partly membranaceous. No skull and no brain.

We need not occupy ourselves with the two last orders, as CYCLOSTOMATA, or Lampreys and Myxines, only include five genera and not a dozen species, while the LEPTOCARDII, or Lancelets, though distributed all over the world, have not more than two species. By many it is not considered a fish at all, but a separate class called ACRANIA, or headless.

Thus according to this system there is little more than a division into —1. fishes with a skeleton of bone, and 2. fishes with a skeleton of cartilage. Before passing to the sub-divisions of these sub-classes it will be well to explain certain anatomical distinctions. Few fishermen will be able to trace them, but to understand them is easily effected and will give great insight into the physiology of fishes.

All fishes (if we exclude the Lancelets) have red blood, and are provided with a complete circulation for the body, another for the gills, and a third for the liver. But fishes differ from other animals in this, that the heart is relatively small, and is provided with only two chambers,

which send the blood on from the veins direct to the gills. Thus all
the veins ultimately discharge their contents into a large chamber of the
heart (*atrium**), whence the blood passes by the pulsation of the ventricle
into the cone or auricle. Valves here prevent its return. The cone or
bulb of the auricle is prolonged into the gill artery, where it soon divides,
sending off a branch to each of the gill arches. After being well revived
by the oxygen of the water, the blood returns from the gills, some to
arteries where a portion is sent off to different parts of the head and
heart, but the main trunk unites to form a great artery which carries
blood to all parts of the muscles, intestines, and tail. In the great
majority of fishes it is only the ventricle which contracts. In sharks,
rays, &c. (Palæichthyes) the auricle also beats or pulsates. It is interesting to study the hearts of fishes, from which much can be learned.
Its simple structure is one that can be easily understood. In the
anterior part will be found the *atrium* with a large *sinus venosus*† or
groove, into which all the veins enter, then the ventricle, and then a
conical hollow swelling at the beginning of the arterial system. In all
the sharks and rays this swelling is still a division of the pulsating
heart. It has a thick muscular layer, but is not separated from the
ventricle by two valves (like the heart of warm-blooded animals) opposite
to each other, yet its interior is provided with many valves arranged in
a transverse series. In Cyclostomata and Teleosteans the enlargement is
a swelling of the artery without a muscular thickening, and it does not
pulsate.

The reference made to the crossing of the optic nerves is worthy of
particular notice. The optic nerves of fishes take their rise from the
optic lobes of the brain. In the Cyclostomata they go straight to the
eye each on its own side. In the Teleosteans they simply cross each
other so that each optic nerve supplies the eye on the other side of the
brain. In sharks and rays the optic nerves unite after leaving the
brain and become merged into one. This compound nerve is cylindrical
for a time, but soon flattens out like a plaited band which can be
separated and expanded.

The blood, as already explained, is carried to the gills. These are the
lungs of fishes, only lungs are organs especially designed to breathe
air, while gills breathe water. Both, however, are for the same purpose,
that is, renewing the blood by the oxygen which both air and water contain, while at the same time it parts with certain elements which it has
removed from the system. In gills this is effected between what are
called the branchial arches. They are usually five in number, of which
four bear gills, while the fifth remains dwarfed.

The gills are fine lamellæ covered with mucous membrane, every one
of which is supplied with a small twig from the branchial artery. This
is subdivided into the finest branchlets, and their termination being very
near the surface is what gives the red colour to the gills. The water
washing over these surfaces while the blood is moved along by the heart,
brings every portion in succession within reach of the oxygen which is
rapidly absorbed. On the inner side of the gills there are sets of organs
called gill-rakers, which act as a sieve to prevent particles of solid matter

\* Latin for court or hall.
† Latin for vein-cavity, or fold.

from passing over the gills. The above account will generally describe the gills of Teleosteans. The sharks and rays have a different arrangement. There are certain other fishes in which the gills are so peculiar, such as some *Siluridæ*, *Labyrinthici*, &c., that larger works on Ichthyology must be consulted concerning them.

Another organ to which attention must be paid is the air-bladder. This is a hollow sac of very varying shape. It is situated in the abdomen, but not within the same sac with the intestines. It is either entirely closed or communicates by a duct with the latter. It is full of gas, not air, and curiously enough the nature of the gas differs in fresh-water fishes and those of the ocean. In the former case the gas is principally nitrogen; in the latter, especially in deep-sea fishes, it is oxygen which predominates. In some fishes the air-bladder assumes the character of a lung; in the *Leptocardii*, *Cyclostomi*, *Chondropterygii* and *Holocephali*, it is wanting: in the others its office is supposed to be useful in altering by compression the specific gravity, and either to change the fish's centre of gravity or enable it to sink and float at will.

The shape of the intestinal tract in fishes is subject to all sorts of variations. As a rule it is shorter in carnivorous fishes and longer in vegetable feeders. It is often uniform throughout, so that the different portions of stomach, œsophagus, intestine, and rectum cannot be distinguished. But this is rarely the case in Teleosteans. There are two forms of stomach common. One is the siphonal, in which the organ is bent into two portions; the other is the cœcal, in which one end is prolonged into a long descending blind sac. Besides this there are commonly among Teleosteans a number of hollow worm-like appendages, varying in numbers in different species from two or three to 200; they open into the intestine at the pyloric* orifice of the stomach, and are called pyloric appendages. They are much used as distinguishing marks between one species and another.

Fishes have liver, spleen, pancreas, and intestines variously modified. The liver is mainly distinguished for the quantity of oil which it contains. Details about these organs are not within the purpose of this work; they are, however, of especial interest to those who wish to pursue the subject. Günther's work on the "Study of Fishes" is recommended for the purpose. What has been already given will be sufficient for all the purposes of classification.† If any intelligent fisherman will take one of the fishes daily met with and examine it so as to see for himself the organs here referred to, he will learn practically what he is taught by reading, and it will effectually be impressed upon his mind.

The skeleton next claims our attention, as it forms a great mark of distinction between the two great sub-classes, Teleosteans and Sharks, &c. The latter were called *Chondropterygians*, on account of the exclusively cartilaginous character of their bones. But there is great diversity in this, for while some have scarcely a consolidated cartilage, others present almost every degree of ossification. In some the vetebræ can be hardly distinguished from one another, in others they are well defined and

* Greek—*Pule*, a gate—the lower opening of the stomach leading to the intestines.

† Those who require the meanings of such terms as pancreas, spleen, &c., or the nature of the organs themselves, can learn all from the shilling manuals of physiology, published by M'Millan, Collins, and others.

complete. In nearly all the *Palæichthyes* there is a peculiar arrangement for the termination of the tail, which is called *heterocercal*. The vertebræ are continued into the upper lobe of the tail, which makes it unsymmetrical. In *Teleostean* the caudal fin is more or less symmetrical on each side of the end of the vertebræ. This is called *homocercal*. In this the caudal fin appears to be centred on the last vertebra, which coalesces with a flat bony plate, the hypural bone, on the flat margin of which the fin rays are placed.

The cartilaginous skeleton of the *Palæichthyes* is peculiar in leaving no sutures in the skull, which is always formed of a single piece, but in which we can distinguish, by means of the projections, grooves, hollows and holes, certain regions which are analogous to the bones in the skulls of other fishes. Again, parts of the vertebræ in certain rays are united into a single body. The gelatinous substance which in other fishes fills the intervals of the vertebræ and communicates from one to the other only by a small hole, forms in several *Chondropterygii* a cord (notochord) which runs through all the bodies of the vertebræ, almost without varying in diameter.

As already stated, most of these details will be useless to the ordinary observer for the purposes of classification. Other portions of the body of a fish must be relied upon. The chief of these are the fins. They are supposed to correspond to the limbs of other vertebrate animals. Those corresponding with the fore limbs are said to be *pectoral*, and those which correspond to the hind limbs are called *ventral*. Vertical fins on the back are said to be *dorsal*, and similar fins underneath, in a line with the axis of the body and near the vent, are called *anal*. The tail fin is called the *caudal*. Any of these may be present or absent; sometimes they are highly developed and extend a great length, or again may be a mere fold in the skin; thus the dorsal fin may extend from the head to the vent, joining the anal and caudal in one continuous line. Many important differences are perceived in the dorsal fin on which systems of classification are founded. The rays which support it may contain spines of bone, or they may be all soft and jointed. The fishes with spinous rayed dorsal fins comprise an immense order called the *Acanthopterygii* or spiny fins. The soft-rayed are called *Malacopterygians*. These spines can nearly always be erected or depressed at the will of the fish. If, when depressed, they cover each other completely so that their points all lie in the same line they are called *homocanth*; but if they are unsymmetrical, that is alternately broader on one side than on the other, they are called *heterocanth*. The anal fin may be divided into one or more fins, or be absent. In *Acanthopterygians* its foremost rays are frequently spinous.

It must be borne in mind also that though the fore and hind limbs are represented by fins in fishes, yet in most cases they are close together and generally near the head. The pectorals or fore limbs with their bony supports are always fixed immediately behind the gill-opening. The ventrals are subject to much variation in position, and formerly were used by zoologists to distinguish large families of fishes by their position. When inserted behind the pectorals on the abdominal surface, the fishes were grouped into one order under the old Linnean arrangement, called *Abdominales*. This included salmon, silurus, pike, mullet, herring, pipe-fish. When the ventral fins were situated on the breast or nearly under the pectorals, they formed an order called *Thoracici* or breast-fish, including dolphins, goby, dory, sole, wrasse, perch, mackerel, gurnard, flying-fish; and if the ventral fins were situated under the throat the fishes were included in an order called *Jugulares* or throat-fish, including cod, whiting, hake, blenny. Fish without ventral fins were classed together as *Apodes* or footless.

It will be seen that this arrangement, however convenient in one respect, took no notice of the resemblances or differences of fishes in many other respects; and thus its groups were of the most mixed description. For this reason it was called artificial, because it gave no knowledge of the harmonious plan which prevails throughout nature.

But while the position of the fins is of little value in classification, the number of rays or spines in the ventral fins is of the greatest importance in grouping the smaller divisions, and for the determination of species; moreover the dorsal and anal rays or spines generally correspond to the number of vertebræ in a certain portion of the backbone; they form therefore constant and unvarying characters by which species, genera, and even families may be distinguished. The only exception to this rule is that if the number of spines is very great a proportionately wide margin must be allowed for variation (Günther). The number of pectoral or caudal rays is rarely of any importance.

A few words more will be necessary to describe certain other parts of fishes which are referred to in systems of classification. The eye is taken as a point for dividing the head into the ante-orbital and post-orbital portions. The organ is proportionately larger in most fishes than in other vertebrates. Fishes with very large eyes are either nocturnal in their habits or live at depths in the ocean to which but little light penetrates. Where scarcely any light is found the fishes have small or rudimentary eyes, or the eyes are hidden under the skin. Fishes inhabiting muddy places have small eyes. The Hon. W. Macleay has described a fish from North Australia (*Polynemus cæcus*) which has a thick membrane over the eyes, rendering it if not blind at least only capable of perceiving light. It was found at the entrance of muddy rivers. This is not an uncommon feature in the genus *Polynemus* all of which have long feelers under the pectoral fins to take the place of eyes.

The space between the eyes is called the inter-orbital space or forehead; that beneath them is called the sub-orbital or infra-orbital. In the ante-orbital space are placed the mouth and nostrils. The mouth is formed by the maxillary or inter-maxillary bones or by the inter-maxillary only in the upper jaw, and by the mandibulary bone in the lower. These bones are sometimes bare, but folds of skin, forming lips, are often added. To the jaws are added the only weapons of attack which fishes have. Sometimes skin appendages called *barbels* are attached to both jaws, which generally are organs of touch. In most fishes the nostrils are a double opening on each side of the upper surface of the snout; they lead into a shallow groove, and in only one family perforate the palate (the Myxinoids). In sharks and rays they are underneath the snout, and more or less confluent.

The gill-cover is called the *operculum*, but this name is only applied to the hind margin. In most Teleosteans there is a semicircular bone in advance of this, with a free margin like a second gill; this is called the *pre-operculum*. It varies considerably in shape, often having a toothed edge or spines, and so becomes useful as a feature in classification. The term *operculum* is, as already said, only applied to the hind margin of the gill-cover, which is divided into movable segments. The under one is called the *sub-operculum*, and the segment above is called the *inter-operculum*; they are separate bones. These bones are sometimes rudimentary and sometimes absent. All of them are frequently referred to in descriptions of fishes in scientific works.

Everyone must have noticed along the sides of most bony fishes a line something like a division between the belly and back. It is sometimes straight, but more often is curved in the most varied manner. This is called the lateral line, and is caused by a series of perforated scales through which mucus is especially secreted, though no doubt all the surface of a fish secretes the same fluid to some extent. This perforated line is provided with abundant nerves, and is called the *muciferous* system. Some fishes have many lateral lines, and our coast and river mullets have none at all.

All fishes are of distinct sex. The females in the majority of instances are oviparous. A few bear the young alive; generally the eggs are deposited by the female and are afterwards fertilized by the male. In the males, lying along the intestines, there are two soft bodies called the milt. In certain seasons these contain a milky secretion, which is the fertilizing fluid. This is deposited on the eggs by the male, or merely injected into the water. A very small quantity of this fluid is sufficient for the impregnation of an immense number of ova, and it is owing to this circumstance that the artificial impregnation and hatching of fishes is easily practised and immense numbers of fish preserved.

In the same relative position as the milt (soft roe) in the male are found the bodies called the hard roe in the female. This is a mass of unimpregnated eggs.

In their young states fishes differ so much from the forms they assume in full growth that they have been very frequently mistaken for distinct species. Though a number of such mistakes have been rectified of late years, there are doubtless still some received species which are only the young of other forms. Until all the stages of growth are known these errors will not be detected.

Having now dealt with all the most necessary elements of scientific classification, we can apply them to the general divisions which are made. Let it be premised, however, that those first entering upon this study must not be disheartened if they find it difficult to identify with certainty some of the details. Though perhaps a little out of place in such a work as this, it may be well to suggest a method by which the details of scientific description may be mastered. Instead of taking an unknown fish and trying to make it out by the aid of a catalogue, such as Günther's or Macleay's, let the student compare a fish, the scientific name of which he is sure of, with the description given of it in these works. When this has been done in the case of ten or a dozen species of different families, the work of identification will be comparatively easy. All who are in reach of the Museum in Sydney, or any of our colonial collections, will find themselves so aided by the named species, that instead of a difficult study the whole process of learning will be a delightful recreation.

In the wood-cut at the end of this chapter, fig. 1 represents—A, the vomerine teeth; B, the palatine; C, teeth on the tongue; D, supra-maxillary; E, maxillary. These terms will be frequently used.

Fig. 2. Gill-cover of *Salmo salar*. A, pre-operculum; B, operculum; C, sub-operculum; D, inter-operculum.

Fig. 3. Gill-cover of Salmon trout; letters the same as above.

Fig. 4. *Salmo salvelinus*. 1, pre-operculum; 2, operculum; 3, sub-operculum; 4, inter-operculum; 5, branchiostegal rays; 6, fixed plates forming immovable posterior margin of the gill-cover; 7, root of pectoral fin.

These figures are adapted from "Freshwater Fishes of Central Europe," by L. Agassiz.

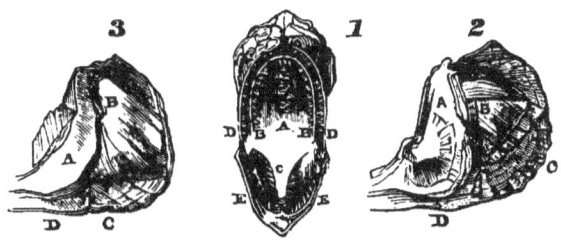

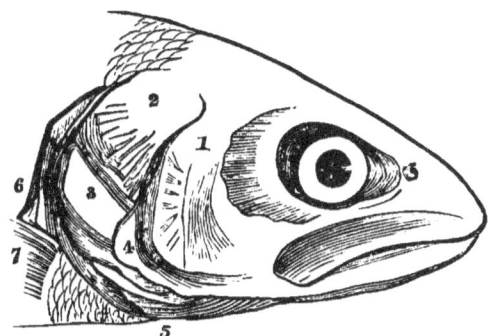

Fig. 1.—Mouth of Trout, *Salmo fario*.
Fig. 2.—Gill cover of *Salmo salar*, Salmon.
Fig. 3.—Ditto of *Salmo trutta*, Sea-trout.
LOWER FIGURE—*Salmo*, head of *Salmo salvelinus* of Central Europe.

## CHAPTER II.
### The Fish Fauna of New South Wales.

In order to understand the characteristics of the fish fauna of the coasts and rivers, it will be necessary to explain something of the fishes of the Australian region generally. They do not differ in any singular or remarkable degree from the fishes of the rest of the world. If there are one or two apparent exceptions to this, it is in the case of some Australian fishes which have representatives, not living, but in remote periods of the world's geological history. Such examples are found in the *Ceratodus*, which inhabits the rivers of Queensland, and the *Cestracion*, or Port Jackson shark (*Heterodontus*). The dentition of the last is extremely like fossil teeth of *Acrodus*, found in mesozoic deposits. *Ceratodus* is an existing ganoid fish, which is abundantly and almost exclusively represented in the Trias formation. Its anatomy also shows it to be a faint connecting link between a lizard and a fish.

Except for such rare instances, the families of Australian fishes are only slightly different from those of other seas. Some are absent and some are very poorly represented, but the great mass of them have relations with these of neighbouring seas or those in which the same conditions of temperature and coast line prevail. There are however minor differences, especially in genera, and these give to Australia whatever distinctive characteristics are possessed by its fish fauna. We find also that these Australian features are more marked on the southern than in the northern coasts. The more remote our coasts are from other lands the more peculiar and distinct are the characters which the coast fishes present, which is just what we might expect. Thus, on the north, north-west, and north-east coasts the fauna is closely connected with that of the Indian and tropical seas, and is in very many species identical with it. The tribes of the colder regions are here wanting and in place we have the fishes of the equatorial zone in all their gorgeous liveries of red, blue, green, and gold, arrayed in those fanciful patterns which awaken the enthusiasm of every naturalist. We find also that as we go southwards on either coast there is a gradual disappearance of the tropical fauna and a mingling of that of the temperate regions. Now if we cast our eyes on the limits of the coasts of New South Wales, we shall find that they lie in regions where the fishes are most likely to be intermediate in character. No part of the Colony is within the tropics, though its northern boundary is not five degrees from the tropic of Capricorn. Again, no part of the Colony abuts upon the south coast, but its southern limits are just at the entrance of Bass's Straits. Thus it is cut off from the equatorial, and again from the South Australian regions. In such a province we can only expect that what is peculiar in its fish fauna will belong to the Pacific Ocean, and in fact it does possess more of the fishes of that area than any other portion of the coasts of Australia. It may be necessary further to mention that these remarks apply to the shore fishes only, that is, fishes which inhabit shallow waters in the neighbourhood of land. Pelagic fishes are those which inhabit the surface or uppermost strata of the open sea and only visit the coasts accidentally in search of prey or periodically for the purpose of spawning. They are not taken into account in this estimate. "They are," says Günther, "subject in their distribution to the influences of light and the temperature of the surface water," but they are independent of the variable local conditions which tie the shore fish to its original home. Deep-sea fishes, or those which inhabit great depths so as not to be influenced by light and temperature, do not come into our estimate at all.

The following are the families of fishes represented more or less numerously in New South Wales, the numbers following being the number of species:—

Percidæ, 50; Squamipinnes, 4; Nandidæ, 2; Mullidæ, 3; Sparidæ, 14; Cirrhitidæ, 6; Scorpænidæ, 11; Teuthididæ, 2; Berycidæ, 3; Kurtidæ, 2; Polynemidæ, 2; Sciænidæ, 2; Xiphiidæ, 1; Trichiuridæ, 1; Acronuridæ, 1; Carangidæ, 15; Cyttidæ, 1; Coryphænidæ, 2; Scombridæ, 10; Trachinidæ, 5; Batrachidæ, 1; Pediculati, 4; Cottidæ, 7; Cataphracti, 1; Gobiidæ, 15; Blenniidæ, 17; Sphyrænidæ, 3; Atherinidæ, 4; Mugilidæ, 7; Fistularidæ, 1; Ophiocephalidæ, 1; Trachypteridæ, 1; Pomacentridæ, 4; Labridæ, 18; Gadopsidæ, 1; Gadidæ. 4; Pleuronectidæ, 9; Siluridæ, 5; Scopelidæ, 6; Salmonidæ, 1; Galaxidæ, 7; Scombresocidæ, 6; Clupeidæ, 12; Chirocentridæ, 1; Symbrachidæ, 1; Murænidæ, 11; Syngnathidæ, 6; Sclerodermi, 21; Gymnodontes, 12; Carcharidæ, 8; Lamnidæ, 1; Scyllidæ, 3; Cestraciontidæ, 2; Spinacidæ, 1; Rhinidæ, 1; Pristiophoridæ, 1; Rhinobatidæ, 2; Trygonidæ, 3; Torpedinidæ, 1; Raiidæ, 1.

Of these 59 families 16 have only 1 species; 8, only 2 species; 5, only 3, and not quite half have more than 3; the largest 50, the next 14, the average about 6.

In the above families there are certain genera and species which are peculiar to the southern temperate zone. This of course includes all the coasts of islands and the continent south of the tropics. This zone is characterized by a striking feature, which is visible in the molluscan and other marine kingdoms, and is also to a certain extent in the land fauna and flora. It is the reappearance of genera and species which are common in the northern temperate zone. The species are said by Günther to be *Chimæra monstrosa, Galeus canis, Acanthias vulgaris, A. blainvilleii, Rhina squatina* (Angel Shark), The John Dorey *(Zeus faber), Lophius piscatorius*, (the Angler-fish) *Engraulis encrassicholus*, (the Anchovy), *Clupea sprattus* (the Sprat), *Conger vulgaris* (Conger eel), *Centriscus scolopax* (Trumpet or Bellows fish), found in Tasmania. It must be said however that some of these instances, such as the Sprat and the Anchovy, are of doubtful occurrence, and there are sufficient differences in the John Dorey, and some of the cartilaginous fishes, as to make many regard them as distinct species.

Instances of genera the same as in the northern temperate zone are still more abundant, and we have the following on the coast of New South Wales. The *Heterodontus*, which is found in Japan and California, but also in the tropics (Amboyna), *Pristiophorus* (saw-sharks, found also in Japan), *Raiia* or thornbacks, *Girella* (black-fish), *Chilodactylus* (the long-finned sea perches, found also on the coasts of China and Japan), *Sebastes* (rock-fish), *Aploactis, Lotella* (rock cod), *Aulopus* (Sergeant Baker).

The following genera are peculiar to the South Australian region, though some, here marked in italics, extend to the coasts of New South Wales:—*Trygonorhina*, *Enoplosus*, *Lanioperca Arripis*, Tephræops, *Trachichthys*, *Chironemus*, Holoxenus, Nemadactylus, *Latris*, Glyptauchen, *Pentaroge*, Anema, Craptolus, *Kathetostoma*, *Leptoscopus*, Platystethus Brachionichthys, Saccarius, *Lepidoblennius*, Patæcus, Acanthoclinus, Diplocrepis, Crepidogaster, Trachelochismus Neophrynichthys, *Labrichthys*, *Odax*, Coridodax, *Olistherops*, Siphonognathus, Pseudophycis, *Lophonectes*\* Brachypleura, *Ammotretis*, *Rhombosolea*, Peltorhamphus, *Teratorhombus*, *Rhomboidichthys*, Chilobranchus, Nannocampus, *Stigmatophora*, *Phyllopteryx*.

Thus out of 44 genera, which are, as far as known, peculiar to the southern portions of the Australian region, New South Wales possesses only 18 of these, numbering in all 23 species.

Of these eighteen genera eight extend to N. Zealand, and one, *Trachichthys*, is found only there and in Tasmania, and one (*Leptoscopus*) has three species in N. Zealand to our one.

The following genera of the equatorial or tropical zone are found on the coasts of N. S. Wales:—*Anthias* 1, *Serranus* 4, *Plectropoma* 4, *Apogon* 2, *Gerres* 3, *Pentapus* 2, *Chrysophrys* 2, *Platycephalus* 3, *Lethrinus* 3, *Trigla* 3, *Sillago* 2, *Sciæna* 1, *Sphyræna* 2, Caranx 4, Psettus 1, Teuthis 2, Pempheris 2, Callionymus 3, *Batrachus* 1, *Petroscirtes* 5, *Fistularia* 1, *Pomacentrus* 1, Heliastes 1, *Synaptura* 2, *Elops* 1, *Chanos* 1, *Chirocentrus* 1, Muræ nesox 1, Muræna 1, *Hippocampus* 1, Monocanthus 17, Ostracion 4, Tetrodon 10, Diodon 4, Carcharias 3, Zygæna 1, *Notidanus* 1, Chiloscyllium 1, *Urolophus*.

All the above thirty genera, including more than 100 species, are characteristic of the tropics, and seldom seen outside the equatorial zone. There are twenty of the same genera marked in italics which are best represented in the tropics. Thus more than one fourth of the fishes found on the coasts of N. S. Wales are tropical. We have only at the most about five and twenty of the exclusively Australian fishes. The rest of our fish fauna is made up of fishes wide in their distribution, or which are restricted to one or two other provinces or peculiar to our coasts. From 160 to 170 species have been found in no other parts of Australia but on the coasts of New South Wales ; but no accurate conclusion can be drawn from this, as our coasts have been so very much better searched than those of any other Colony, and there are very many and very extensive parts of the Australian coast line which have never been searched at all. Thus, for instance, we know nothing of the fish fauna of the coast between Adelaide and King George's Sound, and of the north coast Port Darwin has been the only portion searched, and that far from thoroughly. In a journey I made to the head of the Mitchell I found in some of the tributaries three new species of freshwater fish, including a *Synaptura* or sole. This will show what there is still to be done in the watershed of Carpentaria.

A good many of the 160 species mentioned above are Indian fishes,— that is found on the Indian Archipelago or on the coasts of the Indian Ocean. It may therefore be fairly inferred that, if they are found on the coasts of N. S. Wales, they will also be found on the intervening shores of North and North-east Australia when they are searched.

\* Macleay's *Lophorhombus* is a synonym of this genus.

Part of our fish fauna is made up from the Pacific region, but this is not so large an ingredient as might be expected. Thus we have not many New Zealand species. Dr. Günther is the authority for the statement that many of the *species* of the South Australian province and New Zealand coasts are identical. It is certainly not true of the south-east coasts of Australia, with regard to the *species*, and even the identity of genera is comparatively small, at least of those genera which are characteristic of South Australia or peculiar to its coasts. Of these, as already stated, we have fifteen in N.S.W., only eight of which extend to New Zealand, namely, Trygonorhina, Arripis, Trachichthys, Chironemus, Latris, Leptoscopus, Labrichthys, and Odax.

The other genera which we have in common with New Zealand are—Rhinobatus, Raiia, Trygon, Urolophus, Anthias, Haplodactylus, Pagrus, Scorpis, Trachichthys, Chilodactylus, Scorpæna, Lepidotrigla, Trigla, Bovichthys, Thyrsites, Zeus, Trachurus, Caranx, Cristiceps, Engraulis, Clupea, Conger, Murænichthys, Syngnathus, Monocanthus, Ostracion, Branchiostoma. This makes thirty-five genera common to the south-east of Australia and New Zealand. Nor can we be surprised at this result, as the distance is great, and a very deep sea intervenes between the two provinces. We have not at present the means of estimating the proportion of species which occur also in other islands of the Pacific, but it must be smaller than the New Zealand element, because of the depth of the sea intervening, and the distance separating New South Wales from most of the Pacific Islands.

To sum up the results, therefore, we find that in New South Wales the predominating characteristic of the fish fauna of its coasts is the prevalence of genera peculiar to Australia, but which are more common and better represented on the south coast. The species are for the most part peculiar to New South Wales. Secondly, more than one-fourth of the fauna is made up of tropical genera, about half of the species of which are peculiar to the east coast of Australia. But the data are not sufficient to establish this with certainty. We have furthermore a very few local genera and species. Finally, the rest of the fauna is made up of very wide-spread genera and species. A few of these are European, more from the Pacific, more still common to Australia and New Zealand, but the most having a wide range over the Indian oceans and Chinese seas.

As to our fresh-water fishes, the most of them are peculiar to our rivers, and are not found outside our continent, at least as far as regards the species. A very large proportion belong to the perch family ; and one of the most common, the Murray Cod (*Oligorus*), is not an exclusively fresh-water genus, but has marine representatives on our own coasts and on those of New Zealand. Some other genera are peculiar to Australia, such as *Ctenolates Murrayia, Macquaria, Riverina*; but they are all true perches, and the generic differences are slight. *Therapon* is another percoid genus numerously represented in our rivers, but mostly in the tropics, and it is also known in India. *Lates* is also a perch, and is known in India. We have also numbers of the herring family in the fresh-water streams, and several species of cat-fish or *Siluridæ*. We have also eels, and a peculiar family called *Galaxiadæ*, which is known only elsewhere in New Zealand and the extreme portions of South

America. There is a species in Tasmania (*Galaxias attenuatus*) very closely allied to a *Galaxias* of New South Wales which is identical with one found in the Falkland Islands and South America. We have also in Australia a very singular southern representative of the salmon tribe in our Grayling, which comes nearest to the genus *Coregonus* of Europe. This is *Prototroctes marœna*, but it is not found in New South Wales. All these peculiarities of our river fishes will be specially dealt with in the chapter referring to the fresh-water species.

I now append a list of all the species of fish known to occur in New South Wales either in salt or fresh water :—

### List of Fishes found in the Rivers and on the Coasts of New South Wales.

Class FISHES. Sub-class TELEOSTEI.

Order ACANTHOPTERYGII. Division 1. A. PERCIFORMES.
Family PERCIDÆ.

1. *Lates colonorum*, Günth. "Perch."
2. *L. curtus*, Cast.
3. *L. ramsayi*, Macleay.
4. *Enoplosus armatus*, White. "Old Wife."
5. *Anthias longimanus*, Günth. "Red Perch."
6. *Serranus damelii*, Günth. "Black Rock Cod."
7. *S. guttulatus*, Macleay.
8. *S. undulato-striatus*, Peters.
9. *S. hexagonatus*, C. & V.
10. *Plectropoma annulatum*, Günth.
11. *P. semi-cinctum*, C. & V.
12. *P. susuki*, C. & V.
13. *P. ocellatum*, Günth. "Wirrah."
14. *Genyoroge bengalensis*, Bleek.
15. *Glaucosoma scapulare*, Macleay.
16. *Priacanthus macracanthus*, C. & V.
17. *P. benmebari*, Temm. & Schleg.
18. *Ambassis agassizii*, Günth. (rivers).
19. *Pseudoambassis castelnaui*, Macleay (rivers).
20. *P. ramsayi*, Macleay.
21. *P. jacksoniensis*, Macleay.
22. *Nannoperca australis*, Günth. (river).
23. *N. riverinæ*, Macleay (rivers).
24. *Apogon fasciatus*, White.
25. *A. guntheri*, Cast.
26. *Arripis georgianus*, C. & V.

27. *A. salar*, Richards. " Salmon," N.S.W.
28. *Oligorus macquraiensis*, C. & V. " Murray Cod."
29. *O. mitchelli*, Cast. " Cod." Second species.
30. *Ctenolates ambiguus*, Richardson. " Golden Perch," Rivers.
31. *C. chrystyi*, Cast.   River fishes.
32. *C. flavescens*, Günth.   ,,
33. *Murrayia guntheri*, Cast.   ,,
34. *M. cyprinoides*, Cast.   ,,
35. *M. bramoides*, Cast.   ,,
36. *M. riverina*, Krefft.   ,,
37. *Riverina fluviatilis*, Cast.   ,,
38. *Macquaria australasica*, C. & V.   River fishes.
39. *Therapon cuvieri*, Bleek. " Mado."   ,,
40. *T. richardsonii*, Cast. " Silver Perch."   ,,
41. *T. niger*, Cast.   ,,
42. *T. unicolor*, Günth.   ,,
43. *Agenor modestus*, Cast.
44. *Labotes auctorum*, Günth.
45. *Gerres ovatus*, Günth.
46. *G. subfasciatus*, C. & V.
47. *G. argyreus*, C. & V.
48. *Pentapus setosus*, Bleek.
49. *P. paradiseus*, Günth.
50. *Aphareus roseus*, Cast.

Family SQUAMIPINNES.

51. *Chætodon strigatus*, C. & V.
52. *Chelmo truncatus*, Kaer.
53. *Scatophagus argus*, Linn.
54. *S. multifasciatus*, Richardson.
55. *Scorpis æquipinnis*, Richardson. " The Sweep."
56. *Atypus strigatus*, Günth.

Family NANDIDÆ.

57. *Plesiops bleekeri*, Günth.
58. *Trachinops tæniatus*, Günth.

Family MULLIDÆ.

59. *Upeneoides vlamingii*, C. & V.
60. *Upeneus porosus*, C. & V.
61. *U. signatus*, Günth.

Family SPARIDÆ.

62. *Pachymetopon grande*, Günth.
63. *Girella tricuspidata*, C. & V.
64. *G. simplex*, Richardson.
65. *G. elevata*, Macleay.
66. *G. cyanea*, Macleay.
67. *G. ramsayi*, Macleay.
68. *Haplodactylus lophodon*, Günth.
69. *H. obscurus*, Cast.
70. *Lethrinus nematacanthus*, Bleek.
71. *L. harak*, Forsk.
72. *L. glyphodon*, Günth.
73. *Pagrus unicolor*, C. & V. "Schnapper."
74. *Chrysophrys sarba*, Forsk. "Tarwhine."
75. *C. australis*, Günth. "Black Bream."

Family CIRRHITIDÆ.

76. *Chironemus marmoratus*, Günth.
77. *Chilodactylus vittatus*, Garrett.
78. *C. macropterus*, Richardson. "Morwong."
79. *C. fuscus*, Cast. "Carp."
80. *C. annularis*, Cast.
81. *Latris ciliaris*, Forst.

Family SCORPÆNIDÆ.

82. *Sebastes percoides*, Richardson.
83. *Scorpæna cruenta*, Richardson. "Red Rock Cod."
84. *S. bynœnsis*, Richardson.
85. *S. cardinalis*, Richardson.
86. *Pterois volitans*, Linn.
87. *P. zebra*, C. & V.
88. *Centropogon australis*, White.
89. *C. robustus*, Günth. "Bull-rout."
90. *Pentaroge marmorata*, C. & V. "Fortescue."
91. *Aploactis milesii*, Richardson.
92. *Synancidium horridum*, Linn.

Family TEUTHIDIDÆ.

93. *Teuthis javus*, Linn.
94. *T. nebulosa*, Quoy & G. "Black Trevally."

FISH AND FISHERIES. 17

Division A. BERYCIFORMES. Family BERYCIDÆ.
95. *Monocentris japonicus*, C. & V.
96. *Trachichthys jacksoniensis*, Cast.
97. *Beryx affinis*, Günth. "Nannygai."

Division A. KURTIFORMES. Family KURTIDÆ.
98. *Pempherus compressus*, White.
99. *P. macrolepis*, Macleay.

Division A. POLYNEMIFORMES. Family POLYNEMIDÆ.
100. *Polynemus indicus*, Shaw.
101. *P. macrochir*, Günth.

Division A. SCIÆNIFORMES. Family SCIÆNIDÆ.
102. *Scinæna antarctica*, Cast. "Jew-Fish."
103. *Otolithus atelodus*, Günth. "Teraglin."

Division A. XIPHIIFORMES. Family XIPHIIDÆ.
104. *Histiophorus gladius*, Brouss.

Division A. TRICHIURIFORMES. Family TRICHIURIDÆ.
105. *Trichiurus haumela*, Bl.

Division A. COTTOSCOMBRIFORMES. Family ACRONURIDÆ.
106. *Prionurus microlepidotus*, Lacep.

Family CARANGIDÆ.
107. *Trachurus declivis*, C. & V. "Yellow-tail."
108. *Caranx nobilis*, Macleay.
109. *C. georgianus*, C. & V. "White Trevally."
110. *C. hippos*, Linn.
111. *C. ciliaris*, Bl.
112. *Seriola lalandii*, C. & V. "King-fish."
113. *S. nigrofasciata*, Rüpp.
114. *S. grandis*, Cast.
115. *S. hippos*, Günth. "Samson Fish." New South Wales.
116. ? *Neptomenus travale*, Cast.
117. *Temnodon saltator*, Bl. "The Tailor."
118. *Trachynotus ovatus*, Linn.
119. *T. baillonii*, C. & V.

c

120. *Psettus argenteus*, Linn. "Bat-fish."
121. *Psenes leucurus*, Jenyns.

### Family CYTTIDÆ.

122. *Zeus australis*, Richardson. "John Dory."

### Family CORYPHÆNIDÆ.

123. *Coryphæna punctulata*, C. & V.
124. *Brama raii*, Bl.

### Family SCOMBRIDÆ.

125. *Scomber antarcticus*, Cast. "Mackerel.'
126. *Thynnus affinis*, Cantor.
127. *T. pelamys*, Linn. "Bonito."
128. *Pelamys australis*, Macleay.
129. *Auxis ramsayi*, Cast. "Horse Mackerel," of Sydney.
130. *Cybium commersoni*, Günth.
131. *C. guttatum*, C. & V.
132. *Naucrates ductor*, Linn. "Pilot Fish."
133. *Elacate nigra*, Bl. "King Fish," of West Indies.
134. *Echeneis remora*, Linn. "Sucking Fish."

### Family TRACHINIDÆ.

135. *Leptoscopus macropygus*, Richardson.
136. *Sillago maculata*, Quoy & G. "Whiting."
137. *S. bassensis*, C. & V. "Trumpeter Whiting."
138. *Bovichthys variegatus*, Richardson.
139. *Opisthognatus jacksoniensis*, Macleay.

### Family BATRACHIDÆ.

140. *Batrachus dubius*, White.

### Family PEDICULATI.

141. *Antennarius marmoratus*, Less.
142. *A. striatus*, Shaw.
143. *A. pinniceps*, C. & V.
144. *A. coccineus*, Less & Garn.

### Family COTTINA.

145. *Platycephalus bassensis*, C. & V. "Red Flathead."
146. *P. fuscus*, C. & V. "Flathead."

147. *P. cirronasus*, Richards.
148. *Lepidotrigla papilio*, C. & V.
149. *Trigla pleuracanthica*, Richardson.
150. *T. kumu*, Less & Garn.
151. *T. polyommata*, Richardson. " Flying Gurnet."

Family CATAPHRACTI.

152. *Dactylopterus orientalis*, C. & V.

Division A. GOBIIFORMES. Family GOBIIDÆ.

153. *Gobius bifrenatus*, Kner.
154. *G. semifrenatus*, Macleay.
155. *G. buccatus*, C. & V.
156. *G. flavidus*, Macleay.
157. *G. eristatus*, Macleay.
158. *Eleotris coxii*, Krefft. (River Fish.)
159. *E. grandiceps*, Krefft.
160. *E. compressus*, Krefft.
161. *E. oxycephala*, Schleg.
162. *E. mastersii*, Macleay.
163. *Aristeus fluviatilis*, Cast. Murrumbidgee River.
164. *A. lineatus*, Macleay.
165. *Callionymus calauropomus*, Richardson.
166. *C. calcaratus*, Macleay.
167. *C. latealis*, Macleay.

Division A. BLENNIIFORMES. Family BLENNIIDÆ.

168. *Blennius unicornis*, Cast.
169. *B. castaneus*, Macleay.
170. *Petroscirtes variabilis*, Cast.
171. *P. anglis*, C. & V.
172. *P. solorensis*, Bleek.
173. *P. fasciolatus*, Macleay.
174. *P. guttatus*, Macleay.
175. *P. rotundiceps*, Macleay.
176. *P. cristiceps*, Macleay.
177. *Lepidoblennius geminatus*, Macleay.
178. *Cristiceps nasutus*, Günth.
179. *C. fasciatus*, Macleay.
180. *C. macleayi*, Cast.

181. *C. aurantiacus*, Cast.
182. *C. pictus*, Macleay.
183. *C. argyropleura*, Kner.
184. *Sticharium dorsale*, Günth.

### Division A. MUGILIFORMES. Family SPHYRÆNIDÆ.

185. *Sphyræna novæ-hollandiæ*, Günth.
186. *S. obtusata*, C. & V. "Pike."
187. *Lanioperca mordax*, Günth.

### Family ATHERINIDÆ.

188. *Atherina pinguis*, Lacep. "Hardyhead."
189. *Atherinichthys jacksoniana*, Quoy & Gaimard.
190. *A. duboulayi*, Cast.
191. *Nematocentris nigrans*, Richardson. River fish.

### Family MUGILIDÆ.

192. *Mugil grandis*, Cast.
193. *M. dobula*, Günth.
194. *M. cephalotus*, C. & V.
195. *M. peronii*, C. & V.
196. *M. compressus*, Günth.
197. *M. pettardi*, Cast.
198. *M. crenidens*, Knor.
199. *Myxus elongatus*, Günth.

### * Family FISTULARIDÆ.

200. *Fistularia serrata*, Cuv.

### Family OPHIOCEPHALIDÆ.

201. *Ophiocephalus striatus*, Bl.

### Family TRACHYPTERIDÆ.

202. *Regalecus jacksoniensis*, Ramsay.

## Order II. ACANTHOPTERYGII PHARYNGOGNATHI.

### Family POMACENTRIDÆ.

203. *Pomacentrus dolii*, Macleay.
204. *Parma microlepis*, Günth.

* The other divisions of Acanthopterygians need not be specified.

FISH AND FISHERIES. 21

205. *P. squamipinnis*, Günth.
206. *Heliastes hypsilepis*, Günth.

Family LABRIDÆ.

207. *Trochocopus unicolor*, Günth.
208. *Cossyphus unimaculatus*, Günth.   "Pig-fish."
209. *C. gouldii*, Richardson.   "Blue Groper."
210. *Labrichthys celidota*, Forst.
211. *L. laticlavius*, Richardson.
212. *L. luculenta*, Richardson.
213. *L. güntheri*, Bleek.
214. *L. parila*, Richardson.
215. *L. gymnogenis*, Günth.
216. *L. nigromarginata*, Macleay.
217. *L. dorsalis*, Macleay.
218. *L. labiosa*, Macleay.
219. *L. melanura*, Macleay.
220. *Coris lineolata*, C. & V.
221. *Heteroscarus castelnaui*, Macleay.
222. *Odax balteatus*, C. & V.   "Kelp fish."
223. *O. semifasciatus*, C. & V.   "Rock whiting."
224. *O. brunneus*, Macleay.
225. *Olistherops brunneus*, Macleay.

Order III. ANACANTHINI.
Family GADOPSIDÆ.
226. *Gadopsis marmoratus*, Richardson.   "Black-fish" of rivers.

Family GADIDÆ.
227. *Lotella fuliginosa*, Günth.
228. *L. callarias*, Günth.   "Cod," of Melbourne.
229. *L. marginata*, Macleay.   "Beardy," of N.S.W.
230. *L. grandis*, Ramsay.

Family PLEURONECTIDÆ.
231. *Pseudorhombus russellii*, Gray.   "Flounder."
232. *P. multimaculatus*, Günth.
233. *Teratorhombus excisiceps*, Macleay.
234. *Rhomboidichthys spiniceps*, Macleay.
235. *Solea microcephala*, Günth.
236. *S. macleayana*, Ramsay.

237. *Synaptura nigra*, Macleay.
238. *Plagusia unicolor*, Macleay. "Lemon sole," N.S.W.

### Order IV. PHYSOSTOMI.
#### Family SILURIDÆ.

239. *Copidoglanis tandanus*, Mitchell. "Cat-fish," rivers.
240. *Cnidoglanis megastoma*, Richardson. "Sea cat-fish," N.S.W
241. *C. lepturus*, Günth.
242. *Arius thalassinus*, Rüpp.
243. *A. australis*, Günth.

#### Family SCOPELIDÆ.

244. *Saurus myops*, C. & V.
245. *Saurida nebulosa*, C. & V.
246. *S. australis*, Cast.
247. *S. truculenta*, Macleay.
248. *Aulopus purpurissatus*, Richardson. "Sergeant Baker," N.S.W.
249. *Chloropthalmus nigripinnis*, Günth.

#### Family SALMONIDÆ.

250. *Retropinna Richardsonii*, Gill. Rivers.

#### Family GALAXIIDÆ.

251. *Galaxias krefftii*, Günth. ⎫
252. *G. scriba*, C. & V. ⎬
253. *G. punctatus*, Günth. ⎬ Rivers.
254. *G. coxii*, Macleay. ⎬
255. *G. planiceps*, Macleay. ⎬
256. *G. bong-bong*, Macleay. ⎭
257. *G. nebulosa*, Macleay (marine).

#### Family SCOMBRESOCIDÆ.

258. *Belone ferox*, Günth. "Long Tom," N.S.W.
259. *B. gracilis*, Macleay.
260. *Hemirhamphus intermedius*, Cantor. "Gar-fish," N.S.W.
261. *H. regularis*, Günth. "River gar-fish," N.S.W.
262. *H. argenteus*, Beun.
263. *H. commersonii*, C. & V.

FISH AND FISHERIES. 23

Family CLUPEIDÆ.
264. *Chatoessus richardsoni*, Cast. "Bony bream," of N.S.W. Rivers.
265. *Clupea sagax*, Jenyns. "Herring," N.S.W.
266. *C. sundaica*, Bleek.
267. *C. hypelosoma*, Bleek.
268. *C. moluccensis*, Bleek.
269. *C. novæ-hollandiæ*, C. & V.
270. *C. richmondia*, Macleay.
271. *Etrumeus jacksoniensis*, Macleay.
272. *Elops saurus*, Linn.
273. *Megalops cyprinoides*, Brouss.
274. *Chanos salmoneus*, Bl.

Family CHIROCENTRIDÆ.
275. *Chirocentrus dorab*, Forsk.

Family SYMBRANCHIDÆ.
276. *Chilobranchus rufus*, Macleay.

Family MURÆNIDÆ.
277. *Anguilla reinhardtii*, Steind.
278. *A. australis*, Richardson. "The Eel."
279. *Conger vulgaris*, Cuv.
280. *C. labiatus*, Cast. "Conger Eel."
281. *Murænesox cinereus*, Forsk. "Silver Eel," N.S.W.
282. *Myrophis australis*, Cast.
283. *Murænichthys australis*, Macleay.
284. *Ophichthys serpens*, Linn.
285. *Muræna undulata*, Lacep.
286. *M. picta*, Bl.
287. *M. afra*, Bl. "Green Eel," N.S.W.

Order V. LOPHOBRANCHII.

Family SYNGNATHIDÆ.
288. *Syngnathus margaritifer*, Peters.
289. *S. tigris*, Cast.
290. *Stigmatophora argus*, Richardson.
291. *S. nigra*, Kaup.
292. *Phyllopteryx foliatus*, Shaw.
293. *Hippocampus novæ-hollandiæ*, Steind.

Order VI. PLECTOGNATHI.

Family SCLERODERMI.

294. *Monocanthus hippocrepis*, Quoy & G.
295. *M. convexirostris*, Günth.
296. *M. trachylepis*, Günth.
297. *M. guntheri*, Macleay.
298. *M. spilomelanurus*, Quoy & G.
299. *M. maculosus*, Richardson.
300. *M. castelnaui*, Macleay.
301. *M. freycineti*, Hollard.
302. *M. prasinus*, Cast.
303. *M. margaritifer*, Cast.
304. *M. megalurus*, Richardson.
305. *M. granulatus*, White.
306. *M. rudis*, Richardson.
307. *M. ayraudi*, Quoy & G.   "Leather-jacket," N.S.W.
308. *M. trossulus*, Richardson.
309. *M. oculatus*, Günth.
310. *M. macrurus*, Macleay.
311. *Ostracion concatenatus*, Bl.
312. *O. diaphanus*, Bl.
313. *O. cornutus*, Linn.
314. *O. lenticularis*, Richardson.

Family GYMNODONTES.

315. *Tetrodon lævigatus*.
316. *T. hypselogenion*, Bleek.
317. *T. hamiltoni*, Richardson.
318. *T. virgatus*, Richardson.
319. *T. hispidus*, Linn.
320. *T. firmamentum*, Schleg.
321. *T. lineatus*, Bl.
322. *T. amabilis*, Cast.
323. *Diodon hystrix*, Linn.
324. *D. 9-maculatus*, Cuv.
325. *Dicotylichthys punctulatas*, Kaup.
326. *Orthagoriscus mola*, Linn.   "Sun-fish."

SUB-CLASS II.—PALYICHTHYES.

Order CHONDROPTERYGII.   Family CARCHARIDÆ.

327. *Carcharias macloti*, Mull & Henle.

FISH AND FISHERIES. 25

328. *C. glaucus*, Linn. "Blue Shark."
329. *C. gangeticus*, Mull & Henle.
330. *C. brachyurus*, Günth.
331. *Galeocerdo rayneri*, McD & Barr. "Tiger Shark."
332. *Galeus australis*, Macleay. "School Shark," N.S.W.
333. *Zygæna malleus*, Shaw. "Hammer Shark."
334. *Mustelus antarcticus*, Günth.

Family LAMNIDÆ.

335. *Lamna glauca*, Mull & Henle. "Blue Pointer."
336. *Carcharodon rondeletii*, Mull & Henle. "White Pointer."
337. *Odontaspis americanus*, Mitch. "Grey Nurse."
338. *Alopecias vulpes*, Linn. "Thresher."

Family NOTIDANIDÆ.

339. *Notidanus indicus*, Cuv.

Family SCYLLIDÆ.

340. *Scyllium maculatum*, Bl. "Dog-fish."
341. *Chiloscyllium furvum*, Macleay.
342. *Crossorhinus barbatus*, Linn. "Wobbegong," N.S.W.

Family CESTRACIONTIDÆ.

343. *Heterodontus phillipii*, Lacep.
344. *H. galeatus*, Günth.

Family SPINACIDÆ.

345. *Acanthias megalops*, Macleay.

Family RHINIDÆ.

346. *Rhina squatina*, Linn. "Angel-fish."

Family PRISTIOPHORIDÆ.

347. *Pristiophorus cirratus*, Latham.

Family RHINOBATIDÆ.

348. *Rhinobatus granulatus*, Cuv. "Shovel-nose."
349. *Trygonorhina fasciata*, Mull & Henle. "Fiddler."

Family TORPEDINIDÆ.

350. *Hypnos subnigrum*, Dumeril.

D

Family RAIIDÆ.

351. *Raiia lamprieri*, Richardson. "Thorn Back."

Family TRYGONIDÆ.

352. *Trygon pastinaca*, Linn. "Sting Ray," N.S.W.
353. *T. tuberculata*, Lacep.
354. *Urolophus testaceus*, Mull & Henle.

Family MYLOBATIDÆ.

355. *Mylobatis aquila*, Linn.
356. *M. australis*, Macleay.
357. *Ceratoptera alfredi*, Krefft.

SUB-CLASS LEPTOCARDII.

358. *Branchiostoma Lanceolatum*, Pall, or the Lancelet.
359. *Chætodon strigatus*, Bleek.
360. *Periophthalmus australis*, Castlenau.
361. *Chilodactylus mulhalli*, Macleay.

All the genera that are of any importance or have an economical value are described at sufficient length in the next chapter; but in the foregoing list there are a few genera which are represented in Australia by a great number of species. They are not important as articles of food, but as they are so numerous in our seas, and therefore become characteristic of our fish fauna, a more extended notice of them is desirable. Some remarks will also be added on a few of the remarkable and exceptional fishes which are found on the coasts of New South Wales.

## Frog-fish.

Or *Antennarius* belonging to the order of *Pediculati*, a name which expresses the singular foot-like office of the fins, which are more fitted for walking along the bottom than for swimming. To this order belongs *Lophius piscatorius* or Fishing Frog so well known all over the world. The genus *Antennarius*, of which we have many species in Australia, is distinguished by a very large head and frog-like body without scales, with a peculiar tentacle just above the snout. The species are pelagic, mostly tropical, and found crawling on floating sea-weed in mid-ocean. They cannot swim much, so on the coast conceal themselves amid the stones and sea-weed, holding on by their arm-like fins. They are all highly coloured, yet their hues are assimilated to the surrounding medium, so that it is very difficult to distinguish them in the water. All the species have a wide range, and this arises from their living in the open ocean, attached to sea-weed, whence they may be drifted anywhere. As their colour depends much on the medium where they are found, no doubt a good many different names have been applied to the same fish by naturalists. This is the case with our common *A. marmoratus*, Less., which has a vocabulary of synonyms all to itself. Probably some of our many specific distinctions will be reduced hereafter to two or three forms.

## Gobies or Sea Gudgeons.

These little carnivorous shore fishes give their name to the order to which they belong. They are easily recognized by the peculiar form of the ventral fins, which are united on each side so as to form a circular funnel-shaped cavity. In one genus (*Lepidogaster*), not known in Australia) the united rays of the ventral fins form a flattened round disc like a sucker. The centre of this disc is soft and flexible, so that the fish can use the whole as a sucker, and thus adheres to rocks and stones When attached the heaviest surf will not dislodge them. Hence it is called the sucking-fish or lump fish, but not to be confounded with the *Remora*, the sucking-fish which adheres to the shark. This belongs to the mackerel tribe, and its sucking disc is on its head. The Gobies are pretty equally distributed all over the temperate and tropical coasts, and over 300 species have been described. They prefer rocky coasts, because with their ventral fins they can adhere firmly to a rough surface and defy the force of the waves. One British species lives, breeds, and dies in one year, being like a plant, an annual. In this species also the adult males have long teeth, while those of the female are very small ones. Perhaps some of our own species have these peculiarities, and this is one of the many fields where the Australian naturalists have a fine, easy, and interesting opportunity for observation. On the confines of the northern boundaries of New South Wales may be seen a very remarkable Goby called the "Hopping-fish." The pectoral fins are developed into regular legs, with which the fish hops or leaps along the mud flats with the greatest rapidity. The eyes are on the top of the head, and very prominent, and moreover they can be thrust very far out of their sockets, and moved independently of one another*, thus the fish can see long distances around, and overtake the small crabs in spite of the long stalks to their optics. This fish is called *Periophthalmus australis.* Cast. I have not included it in our list, for it is a tropical form, yet it is said to be found on the mud-flats of the Richmond River.

*Callionymus* or "Dragonets" are also gobies, of which we have three species, all beautifully coloured marine fishes. In almost all the species the mature males have the fin-rays prolonged into filaments, and the fin membranes brightly ornamented.

## Blennies.

We have in Australia a good number of these fishes, which are small littoral forms, abundantly distributed on tropical and temperate coasts of all countries. One of the principal characters of the family is the ventral fin, which instead of being a prominent member, as amongst gobies, is rudimentary or absent, and in any case has never more than four rays. These fins are constantly jugular or in the throat, and either have no function or become prongs, by means of which the fish move rapidly along the bottom. The teeth are prominent, and in the Wolf-fish (*Anarrhicas lupus*), which attains a length of 6 feet, they are really formidable.

* This peculiarity they share with many blennies, pipe-fish, and sea-horses *Hippocampi*).

Of the genus *Blennius* we have only two species, but there are six of *Petroscirtes*, which is a fish without scales, and a long single spinous dorsal fin. The teeth are also a long single series, with a strong curved canine tooth behind. The gill opening is reduced to a small slit above the root of the pectoral fin, and the ventrals are curved organs of two or three strong rays. They are all small species, but look formidable. They frequent the pools of rocks.

*Cristiceps* is another kind of Blenny, of which there are many species in Australia, and six in New South Wales. In this genus there are two dorsal fins, and the ventrals have one long spine, with two or three rays. The gill opening is also wide. One remarkable feature about these fishes, which are often most brilliantly coloured with tints of bright green, violet, purple, yellow, and carmine, is that they bring forth their young alive. The young fishes are admirable objects for seeing the circulation of the blood under the microscope.

## The Wrasses.

Some of our Wrasses or *Labridæ* will be described elsewhere, as they are useful as food fishes, but there is one genus, *Labrichthys*, which is so numerously represented that, though not much caught as food fish, it deserves a special notice. It has the characters of the family, but the body is compressed, covered with large scales, and a more or less pointed snout. The opercles are scaly, and the cheeks more or less so, while the pre-opercle is not serrated, and the lateral line is continuous. The teeth are in a single series, but sometimes an interior line, and generally a canine tooth behind. There are nine spines and eleven rays in the dorsal fin and three and ten in the anal. The difference between these fishes and the Gropers (*Cossyphus*) is that the latter have four anterior canine teeth in each jaw. They are all brightly-coloured fish, but not growing to any size. We have no less than twenty-seven species in Australian waters, of which one-third are caught upon our coasts.

## Scopelidæ.

The family is remarkable for containing fishes which have luminous glands upon them for giving light to their path in the deep. We have none of this genus (*Scopelus*), but other genera of the family, which have the most awful-looking teeth that a fish can possess. If any one will turn to page 586 of Günther's Study of Fishes he will see what is meant, and what kind of an animal this fish must be when 6 feet long. It is called *Plagyodus ferox*. It has been caught off Tasmania, and very probably will be found off our coasts. At page 42 of the same work there is a portrait of another unamiable-looking member of the same family. This is a *Saurus*, of which we have one species, and three of *Saurida*, a closely allied genus, having a few more teeth. Fortunately they are not large.

## Sea-horses.

This is a name applied to the genus *Hippocampus*. These strange fishes are known in Europe as well as in Australia, and derive their name from the resemblance of the head and fore-part of the body to that of a horse. They are mostly tropical, and they belong to an order which have the gills laminated, but composed of small rounded lobes attached to the branchial arches. The gill-cover is a large simple plate. One wonderful peculiarity in the genus *Hippocampus* is that the males carry the eggs in a sac at the base of the tail, opening near the vent. The body is divided into regular rings and transverse ridges, and where these cross each other, the tough, leathery skin has tubercles or points. The tail is square and apparently rigid, but it easily curls up and seizes hold of any object, by means of which it anchors itself. When swimming about the Sea-horse keeps an upright position, but the tail is ready to grasp any object it meets in the water. It quickly entwines in any direction round weeds or other objects, and darts at its prey with great quickness. When the pectoral fins are large and expanded, so as to be like wings, then the Sea-horses are said to belong to another order, *Pegasidæ*, or Flying-horses, of which we have two species in Australia (*Pegasus natans*, Moreton Bay and Torres Straits, and *P. lancifer* in Tasmania), but none in New South Wales.

## The Phyllopteryx.

But of all the curious fishes that ever were seen *Phyllopteryx* is the most extraordinary. It is the ghost of a sea-horse, with its winding-sheet all in ribbons around it; and even as a ghost it seems in the very last stage of emaciation, literally all skin and grief. The process of development by which this fish attained to such a state must be the most miserable chapter in the history of "natural selection." If this be the "survival of the fittest," it is easy to understand what has become of the rest. Natural selection must have inflicted upon the family harder terms than those which were imposed on Count Ugolino by his enemies. There is a good likeness of one species in Günther's Study of Fishes, p. 682. Never did the famishing spectres of the ancient mariner's experience present such painful spectacles. If these creatures be horses, they must be the lineal descendants of those which were trained to live on nothing, but unfortunately perished ere the experiment had quite concluded. The odd things about these strange fishes is that their tattered cerements are like in shape and colour to the sea-weed they frequent, so that they hide and feed with safety. Thus the long ends of ribs which seem to poke through the skin to excite our compassion are really "protective resemblances," and serve to allure the prey more effectually within reach of these awful ghouls. The *Phyllopteryx* is therefore, in spite of his rags and emaciation, an impostor, and like many a sturdy human beggar puts on the aspect of misery more effectually to ply his trade. The appendages to the spines are well worth a study. Just as the leaf-insect is imitative of a leaf, and the staff insect of a twig, so here is a fish like a bunch of sea-weed. If this is development, it stopped here only just in time; one step more and it would have been a bunch of kelp.

## CHAPTER III.
## Our Marine Food Fishes.

THE study of fishes, were it only in an economical point of view, is of the greatest importance and interest. As articles of food, though but one of the uses of fishes, they must even command our attention, yet their value in this respect is hardly appreciated. Fish is known to be a light and easily digested diet, but it is not known that it contains as much *protein* as pork, and consequently 100 lbs. of fish contains as much nourishment as 200 lbs. of wheaten bread and 700 lbs. of potatoes. This encouraging fact may well awaken our interest sufficiently to enable us to get over the following dry technicalities which are a necessary key to the comprehension of the subject.

It will be borne in mind that fishes are divided into four sub-classes, viz. :—*Teleostei, Palæichthyes, Cyclostomata,* and *Leptocardii.* Each of these sub-classes are divided into orders and sub-orders. Each of the orders are separated into divisions, and these again into families. The families are made up of similar genera, and the genera comprise one or more distinct species. It is very difficult for the mind to grasp these distinctions at once, because their number becomes so bewildering. It will facilitate the comprehension of these methods if we suppose the class fishes to represent a kingdom ; the sub-classes, the counties ; the divisions, the parishes ; the families, the villages or towns ; the genera, the houses ; and the residents, the species.

### Sub-class TELEOSTEI.

Heart with a non-contractile arterial bulb, optic nerves crossing, intestine without spiral valve, skeleton bony, vertebræ completely formed, tail generally homocercal.

This sub-class is divided into six orders :—

1. *Acanthopterygii.*—Part of the rays of the dorsal, anal, and ventral fins composed of non-articulated spines. The lower pharyngeal bones separate. The air-bladder in the adult without a pneumatic duct.

2. *Acanthopterygii pharyngognathi.*—The same features as the last, but the pharyngeal bones united.

3. *Anacanthini.*—Vertical and ventral fins without spinous rays. Ventral fins (if present) jugular or thoracic. Air-bladder (if present) without pneumatic duct, lower pharyngeal separate.

4. *Physostomi.*—All the fin rays articulated ; only the first of the dorsal and pectoral fins is sometimes ossified. Ventral fins if (present) abdominal and without spine. Air-bladder (if present) with a pneumatic duct.

5. *Lophobranchii.*—Gills not laminated, but composed of small rounded lobes attached to the branchial arches. Gill cover reduced to a large simple plate. A dermal skeleton replaces the more or less soft integuments.

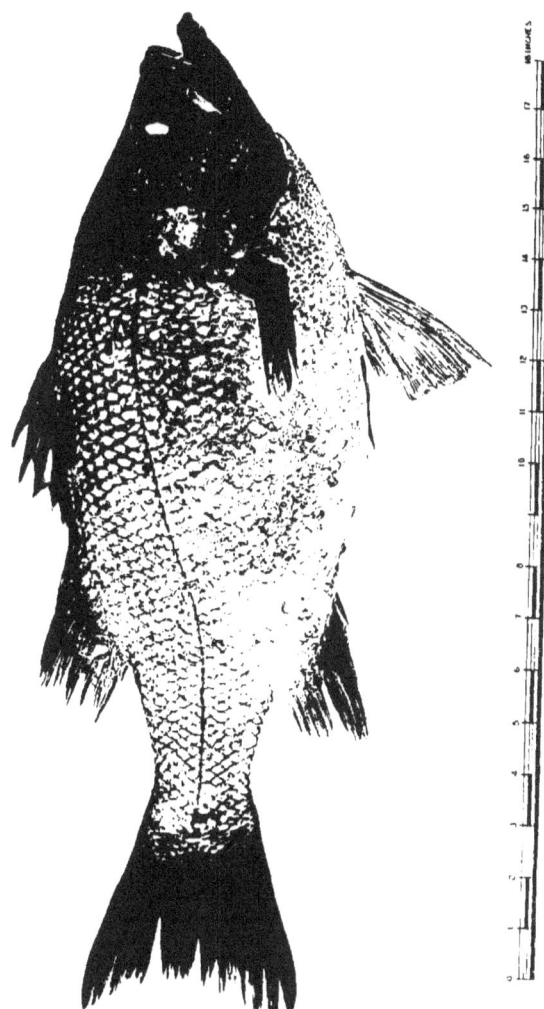

6. *Plectognathi.*—A soft dorsal fin opposite to the anal; sometimes the elements of a spinous dorsal. Ventral fins none, or reduced to spines. Gills pectinate. Air-bladder without a pneumatic duct. Skin with rough plates or shields, or with spines, or naked.

Such a large number of existing fishes are included in the first order that no less than nineteen divisions are made by Dr. Günther. As, however, only a portion of these are found in Australia, it will be better to proceed at once to the characters of the families, or at least such as New South Wales is interested in. The first which occupies our attention is the family of Percidæ or Perches, which is thus characterized :—

First Division.—*Acanthopterygii.*—*Perciformes.*—Body more or less compressed, elevated, or oblong, but not elongate ; vent remote from the extremity of the tail, behind the ventral fins if they are present. No prominent anal papillæ. Dorsal fin or fins occupying greater portion of the back ; spinous dorsal well developed, generally with stiff spines of moderate extent, rather longer than or as long as the soft ; the soft anal similar to the soft dorsal, of moderate extent or rather short. Ventrals thoracic, with one spine and four or five rays.

These features are shared by ten families, of which five at least are important to the fisheries of New South Wales.*

## I.—Fam. PERCIDÆ.

Scales rarely extending over vertical fins, lateral line generally present, continous from the head to the caudal. All the teeth simple and conical, no barbels. No bony stay for the operculum.

This family is very large, and consists of carnivorous fishes, which are distributed all over the world. Fossil genera belonging to Australia are found in European tertiary formations, such as *Lates, Dules, Serranus,* and it is not improbable that they may be found in this Colony also. One of our commonest fishes is *Lates colonorum,* Günth. The genus is distinguished by a compressed body, viliform teeth, teeth on the palatine bones, none on the tongue, no canines. Two dorsal fins, the first with seven or eight, the anal with two spines. Preoperculum with strong spines at the angle of the lower limb, preorbital strongly serrated.

### The Perch.

(Plate I.)

*Lates colonorum,* the perch of the colonists, is easily distinguished by its continuous dorsal fin, which according to Macleay should remove it to another genus. The fourth spine in the dorsal fin is the longest, third anal spine a little longer than the second, lateral line slightly curved. Colour silvery, darker on the back.

* It must here be mentioned that the arrangement followed in the essay is that of Günther's latest work on the Study of Fishes, and differs in the grouping from that adopted by the Blue Book Report of the Commissioners on our Fisheries. The difference, however, is very trifling.

The perch (*Lates colonorum*) is really a fresh-water fish, but as it is often brought to the Sydney market from Broken Bay and other salt water estuaries after freshes in the rivers, we include it among the coast fishes. It is a very delicious fish, but never attains a great size, and is perhaps of more value for the sport it affords to the amateur fisherman than as an article of food.—R.R.C.*

The perch affords good sport to anglers. It loves quiet, shady, and deep holes in the rivers, but when the tide is flowing it may be caught in the stream. It is very voracious. In winter the bait is a small mullet or herring, or better still one of the large grubs that bores into trees. In the early spring months it will take a moth readily, either sunk or on the surface. The artificial salmon-fly is also a splendid bait for trolling at this time. When moths get scarce a frog is a good bait at night. It must be fastened so that it can swim, or if dead, must be played on the water to imitate a frog swimming. No perch can resist that bait at night. In summer grasshoppers, especially that known as the "Percher," a red species, are good bait, but the best is a black house-cricket or an earthworm. This is a very attractive bait, and if the perch are in a pool, the lines are no sooner down than the bait is taken. For the rest of the year a prawn is the best bait, that is when crickets cannot be got. The bait should be at least 4 feet from the float. In landing the fish great care should be used, as the mouth is weak and is easily torn away. The fishes run from 1 lb. to 7½ lbs. The largest are caught in the holes of tributary streams rather than in the main river. The Hunter River is much frequented by anglers for this species of *Lates*.

We have eight other species of *Lates*, two others being found in New South Wales. The perch of the Ganges and other East Indian rivers (*L. calcarifer*) enters freely into brackish water, and extends to the rivers of Queensland. F. Hamilton, in his Account of the Fishes found in the River Ganges, &c. (Edin., 1822, 2 vols., 4to), says of this fish that the vulgar English of Calcutta call it "Cock-up," and that it is one of the lightest and most esteemed food brought to table. Salt water specimens 2 feet in length are the best.

## The Old Wife.
(Plate II.)

The "old wife" (*Enoplosus armatus*, White) is another fish which from its small size is not esteemed nearly so highly as it ought to be. It is a most exquisite fish. It is caught only in the seine net, and never in great quantity, but it is found at all seasons, both young and adult, in Port Jackson and all the harbours of the coast.—R.R.C.

The genus *Enoplosus* is distinguished by a much elevated body, the depth being still more increased by high vertical fins. All the teeth are viliform, without canines, and are on the tongue as well as all the palate bones. Two dorsal fins, the first with seven spines. Preoperculum serrated with spinous teeth at the angles. Scales of moderate size.

The species of our coast is the only one of the genus known in N. S. Wales, and it is easily distinguished by the very elevated body, with eight black transverse bands on a whitish ground.

* The initials R.R.C. stand for Report of the Royal Commission, which will be constantly referred to.

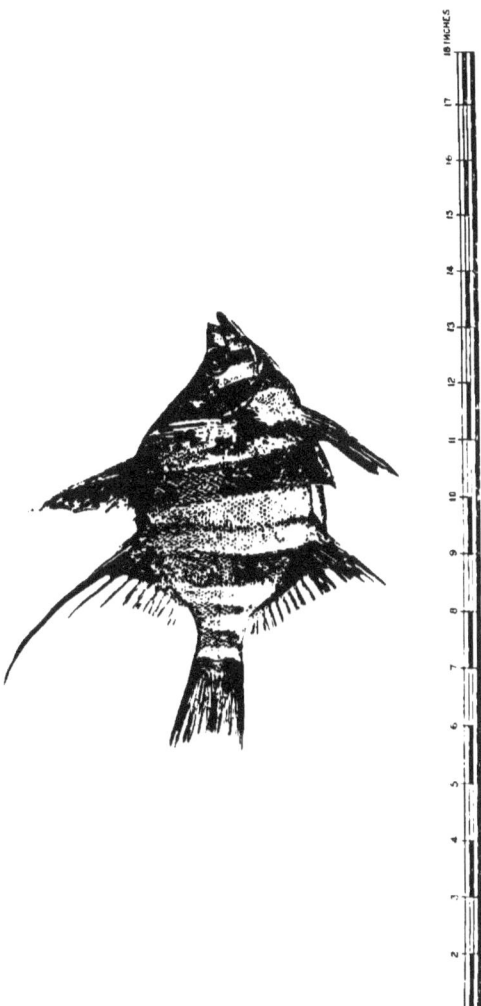

*Enoplosus armatus.*—WHITE.

THE OLD WIFE.

## The Longfin.
(Plate III.)

*Anthias longimanus*, Günth. is a good fish that finds its way to the market occasionally, though probably so rarely that it is not known by any local name. The genus is known by a rather short compressed body, with scales of moderate size. Teeth villiform on all the palatine bones, with small canines in both jaws. One dorsal fin, with generally ten spines, anal fin with three, tail forked. The rays of the fins sometimes prolonged. The species figured may be known by its uniform red colour and the great length of the pectoral fins. All the fins are nearly covered with scales. There are two other species known in Australian waters, which like all the members of the genus are beautifully coloured, the prevailing tints being pink and yellow. Aristotle says that fishers of sponges called it sacred, because no voracious fishes came to the places which it frequented, and the diver might descend with safety. (G.S.F.*)

## The Rock Cod.

*Serranus* is a genus with oblong compressed body and small scales. Teeth villiform on the vomer and palatine bones, none on tongue. Very distinct canines in both jaws. One dorsal, mostly nine or eleven short spines, rarely eight, ten, or twelve; anal fin with three. Pre-operculum serrated behind, and at the angle, but not below. (Günth.)

These are commonly called "sea perches." A few enter brackish and even fresh water, one having been found high up the Ganges, but all spawn in the sea. There are very many varieties known, probably 150, but they vary so much that specific distinctions are extremely difficult to define. Many are most agreeably coloured, with spots, cross-bands, and stripes. These fishes are small, but some reach a length of 3 or 4 feet, and become dangerous to man. Instances of bathers having been attacked by a gigantic species, not uncommon at the Seychelles and Aden, are on record, where death resulted from the injuries received. All the species are eatable (G.S.F.) In New South Wales the best marketable species, *S. damelii*, Günth., is distinguished by being of an entire purplish black, with generally a black spot on the base of the tail at the end of the soft dorsal; end of tail black with a white fringe. In fresh specimens there are faint traces of blue spots; body elongate; height about a third of length without the caudal fin; canine teeth very strong; preoperculum very finely serrated behind; the central spine of the operculum strong; scales very small; dorsal formed of eleven spines, the two first lower than the third, the others becoming rather shorter as they extend backwards, soft parts of fourteen rays rather higher than the spines; caudal fin rounded; anal with three spines, of which the first is shorter and the others almost equal. Extreme size about 3 feet,

The genus *Serranus* comprises most of the fishes known as "rock cod." There are many species of it in these seas, and the number increases in the warmer latitudes of the north, but one only is sufficiently useful as an article of food to merit notice, and that is the "black rock cod" (*Serranus damelii*, Gunther), without exception the very best of all our fishes. It is found on all the rocky parts of the coast, and in the harbours about bold headlands. It takes the hook

* Günther on the Study of Fishes, Edinb., 1880.

readily, and is never captured by the net. It attains a great size, fishes weighing 35 and even 40 lbs. being not uncommon. It is found as far as Jervis Bay to the south; to the north it has probably a much more extended range. It has been observed of this fish that those caught off the "Solitaries" and other places to the north of Port Jackson are as a rule of larger size than those found to the south. It is rarely seen now in the Sydney market, owing to the great falling off in the supply obtained from Port Jackson Heads, Coogee, and other places where it was formerly abundant. It is difficult to account for this diminution of the supply, for the causes which undoubtedly affect the supply of other fishes do not apply to this. The spawn is not deposited in shallow bays constantly raked by nets, the young are never taken in the seine, and the number of the adult fish captured has never been sufficient to account for the deficiency. It is probable that the fish has merely sought retreats further removed from the stir and traffic of Port Jackson.—R.R.C.

## The Wirrah.

(Plate IV.)

"WIRRAH" or PLECTROPOMA is a genus similar to that of Serranus, but armed with a row of spinous teeth on the lower jaw which are directed forwards, besides the pair of canines above. The dorsal fin has from seven to thirteen spines. This is a tropical fish for the most part. There are about thirty species known, of which we have a dozen in Australian seas. Our common marketable species is *P. ocellatum*, Günth. (*P. cyaneo-stigma* in R.R.C.) In this species the body is equal to the length of the head, and two-fifths of the whole without the caudal. Preoperculum with three spines, beneath the anterior of which is the strongest spine, which is sometimes bifid. Colour, brownish; head, body, and base of the fins with numerous roundish spots, bluish in the centre and black round the margin, more numerous and smaller in older fishes.

One of our most experienced amateur fishermen (Mr. A. Oliver) informs me that the wirrah is often mistaken by the tyro for *Serranus damelii*. They are both percoid fishes; but here the resemblance ends. A good black rock cod is equal to the best turbot, and the best wirrah has the flavour and consistency of leather, which no sauce or cooking can change. Both fishes are remarkably tenacious of life. They are lively in the boat or basket long after every other fish has ceased to move.

The genus *Plectropoma* is also numerously represented in our seas; it does not however furnish the market with a single species of value as food. The best known species is the "wirrah" of the fishermen—*Plectropoma cyaneo-stigma* of Günther.—R.R.C.

## The Glaucosoma.

*Glaucosoma scapulare*, Macleay. An excellent food fish, but so rare that it need not be described, especially as it is figured at the end of the first volume of Macleay's catalogue. The name of Ramsay as the authority, there given, should probably be Macleay, as its first published description was by Macleay, who quotes the name as a manuscript one of Ramsay's. See Proc. Linn. Soc. N. S. Wales, vol. v, p. 334. It is sometimes called a Jew-fish, because another species, *G. hebraicum*, goes by that name in Western Australia.

PLATE III.

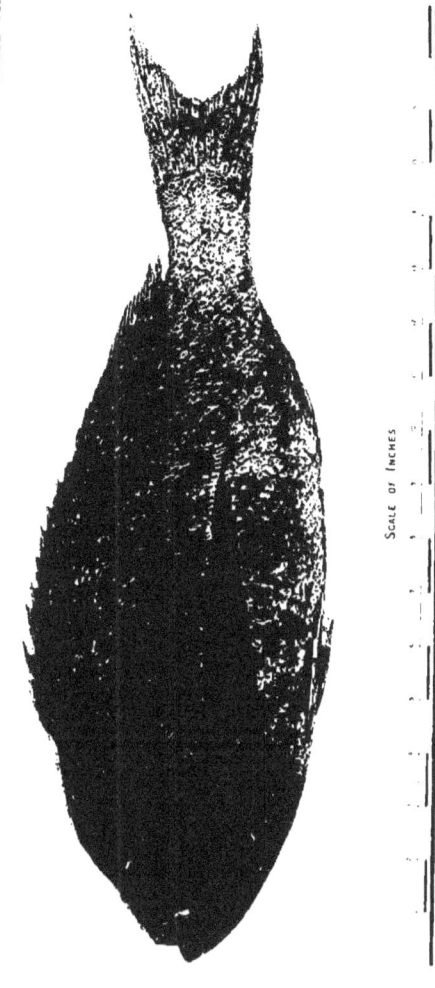

SCALE OF INCHES

*Anthias longimanus.*—GÜNTH.

LONGFIN.

## The Bull's-eye.

PRIACANTHUS. A well-marked genus, easily recognized without direct relation to other percoid genera. The body is short, compressed, covered with small rough scales, which extend also over the short snout. Lower jaw and chin prominent. Eye, large. All the teeth villiform, and present on the vomer and palatine bones. One dorsal with ten spines, anal with three. Pre-operculum serrated, with a more or less flat triangular spine at the angle. (G.S.F.)

The "bull's-eye" of the fishermen—*Priacanthus macracanthus* of Cuv. and Val., is a fairly good edible fish. It comes into the harbour in considerable shoals in the latter end of summer or beginning of winter, some occasionally appearing as early as January. Its visits are probably made for spawning purposes, but we have no direct evidence to that effect, and we cannot find that the young fry are ever seen. The scarcity and irregularity of the supply causes this fish to be little cared for by the dealers. It is frequently mistaken for the "nannygai."—R.R.C.

This species may be known by its large eye, which is more than one-third of the diameter of the head. The ventral fins are long. The colour is of a light silvery grey with a pinkish tint. The head is pink and the belly silvery. Fins of a reddish pink, ventrals red, the back part of the dorsal, anal, and ventral fins having two rounded dark spots on each membrane; end of the caudal rather dark. Length, eight to twelve inches. Mouth extensible. In some respects this fish might be confounded with *Anthias longimanus* by inexperienced observers, but the colour and the eye will easily distinguish it.

## The Salmon.

(Plate V.)

ARRIPIS. Body oblong, covered with scales of a moderate size. All the teeth villiform, without canines. Teeth on the vomer and palatine bones. One dorsal fin with nine slender spines, anal with three, pre-operculum denticulated.

*Arripis salar.* Günth. Cat. Fish. Is in the adult state the salmon of the Australian fishermen, and their salmon trout is the young. It is of a greenish lead colour, with the upper part of the head a deep black; on the upper part of the body are numerous irregular black spots. The operculum and the end of the pectorals are usually tinged with yellow. Its length is sometimes over 22 inches. The young specimens are of an olive green on the upper parts, with the sides and lower parts of a silvery white. On the sides and upper surface extend three or four longitudinal lines of rather large rounded and golden spots, numbering from fourteen to nineteen on each line. Dorsal fin transparent and bordered with black, caudal yellow, with its terminal part black, anal and ventrals white, pectorals yellow, sides of the head and eye of a bright yellow. This is the most common of all Victorian fishes, and the young only take the adult livery when they are at least a foot long. During the cold months of winter the adults are hardly ever seen, but they are common in the summer. In Victoria and South Australia several distinct cases of fish-poisoning have been traced to this species.

By many it is said that if eaten perfectly fresh there is no danger in making use of it, as it is one of the most abundant food fishes in the Colony of Victoria. It is sold in great quantities by the hawkers round the suburbs of Melbourne. At best it is but a poor fish for the table, yet, strange to say, there is considerable difference in this respect between what is caught in Port Phillip and on our seaboard: with us it is considered to be one of the worst of food fishes and scarcely palatable. Prof. McCoy* says, of the Victorian *Arripis* :—"Nearly all the cases of fish-poisoning in Victoria are referable to this species. Some persons are under the impression that the bad consequences are due to incipient decomposition; but I am certain that this is not always the case, as I have known several instances in which the effects were strongly marked after eating perfectly fresh examples, caught only an hour or so before cooking. It is curious that only at certain times, and to certain people, that this fish is more or less poisonous, while certainly good for food under other circumstances not yet understood. I have known three out of five people made seriously ill from eating at breakfast newly caught fish from one basket, and the two others felt no inconvenience whatever. The symptoms are generally a few hours after eating, an extraordinary redness or flush of the skin, particularly of the face, often followed by an eruption, which soon passes away, with great derangement of the digestive organs, headache, vomiting, &c. Some cases of death have been reported, but generally the bad symptoms pass away in a few hours or days. Dr. Youl, the city Coroner for many years, informs me that though he has seen many of these cases of fish-poisoning, the deaths reported were found by the Jury to be due to other causes. The flesh has often a dull pinkish tinge, which may be one of the reasons for the popular application of the names of 'salmon' and 'salmon trout' to this fish, which does not resemble the true salmon in any important respect." It seems to "school" about the latter end of summer, when shoals of astonishing magnitude annually visit our shores. It is the *A. truttaceus* of Cuvier and Valenciennes, and is on the whole a large and beautiful fish.

Mr. Macleay is of opinion that its evil reputation has arisen from the rapidity with which it decomposes after capture. It is said to commence to spawn in September on the east coast. The fry are unknown, as the young are developed into the so-called trout form when they come into the bay. They are caught with the net and line. There is much variation in the colours of the young.

Of late years some fishermen of our port have been trying to devise a drift or purse net for this fish, which sometimes commands as much as 4s. a dozen. They have not as yet succeeded.

In this family (*Percidæ*) are included some of our fresh-water fishes. It will however be more convenient for the purposes of this volume if they are treated altogether in a subsequent chapter.

* Prodromus of the Zoology of Victoria. Decade II., p. 22.

*Plectropoma ocellatum.*—GÜNTH.

THE WIRRAH.

## II.—Fam. ACANTHOPTERYGII PERCIFOMES—SQUAMIPINNES.

Body compressed and elevated, covered with scales, either finely ctenoid or smooth. Lateral line continuous, but not extended over the caudal fin. Mouth in front of the snout, generally small, with lateral cleft; teeth villiform or setiform, in bands without canines or incisors. Dorsal fin consisting of a spinous and soft portion of nearly equal development, anal with three or four spines, developed similarly to the soft dorsal, both many-rayed. Vertical fins more or less densely covered with small scales. Lower rays of pectoral branched, not enlarged, ventrals thoracic with one spine and five soft rays. Stomach cœcal. Eye lateral, of moderate size.

The name of this family will give an easy clue to its identification. It means scaly-finned. The soft and frequently the spiny portions of the dorsal and anal fins are so densely covered with scales that the boundary between where the fin begins and the body ends is quite obscured. The species are generally small, and distinguished for the extraordinary variety and beauty of their colours, including such well-known forms as the *Chætodons*, of which two are found in Port Jackson.

It is said that all the family are carnivorous, and the "Sweep" is no exception. It can be caught with almost any bait. Mr. Oliver says, that he has caught thousands with a live or meat bait. But the odd thing in Sweep fishing is that in nine cases out of ten they are hooked foul. The neighbourhood of coral reefs abounds in forms of this beautiful group.

## The Sweep.

(Plate VI.)

This is one of a very numerous tropical family of fishes, remarkable for eccentricity of form and variety of marking, but the temperate coast of New South Wales can boast of only a few species, and of these one only, the "sweep" (*Scorpis æquipennis*), can claim recognition among our edible fishes. It is not much thought of, yet at times it is brought to market in considerable quantities, and finds consumers at fair prices. The schooling season is midsummer, and the spawn is probably deposited in the harbour, as the young sweeps are frequently caught in the seine. The air-bladder is said to be large in this fish, so that it may be found to be valuable as an isinglass producer. It is seldom caught except in the seine, and is probably entirely a vegetable feeder.—R.R.C.

The species thus referred to is described as having the family characters, with a moderate snout. Dorsal fin with nine or ten spines, anal with three, and the soft portions of both densely scaly. Jaws with an outer series of stronger teeth: teeth also on the vomer and palatine bones. Seven branchiostegals, air-bladder present, pyloric appendages very numerous. Lower margin of the preoperculum finely serrated. The dorsal and anal fins not falcate, the rays becoming shorter posteriorly. Colour uniform brownish black.

## III.—Fam. MULLIDÆ.

Body rather low, slightly compressed, covered with thin scales, with or without extremely fine serrations; two long movable barbels. Lateral line continuous. Mouth, like the last family; cleft rather short; teeth very feeble. Eye like the last. Two short dorsal fins remote from one another, the first with feeble spines; anal like the second dorsal, ventrals with one spine and five rays. Pectorals short, Branchiostegals four, stomach siphonal.

This family is known as "Red Mullets." They are marine fishes, but many species enter into brackish water. They are more tropical than temperate, but extend into both seas. None attain to a large size, but all are very highly esteemed as food. They are caught with the net. We have in Port Jackson *Upeneoides vlamingii*, Cuv. and Val., a red fish with a violet spot on each scale, and violet oblique streaks on the cheeks; and *Upeneus porosus*, a red fish with two silvery streaks between the eye and the mouth, parts above the lateral line darker, and the spinous dorsal blackish. The differences between these two genera is that *Upeneoides* has teeth on both jaws, on the vomer and palatine bones, and *Upeneus* has teeth in both jaws in a single series and none on the palate. The name of this family is a source of much confusion. It is derived from the Latin word *Mullus*, which in the form of "Mullet" we apply to the well-known fishes of quite a different family, the *Mugilidæ*. Another fish to which the term "Red Mullet" is applied is of the family *Cottidæ* or Gurnards. The Greek for these fishes is "*Trigle*," which Oppian derives from their breeding thrice a year. The Italian name is still "*Triglia*." An extraordinary value was set on these fish by the Romans in the time of the Cæsars. This was *Mullus barbatus*. Pliny, Seneca, Horace, Juvenal bear witness to the extravagance of the wealthy of those times with regard to this fish. Nothing was considered more entertaining than to watch the change of its beautiful colours when expiring, and then when dressed it was the grand dish of the feast.

## IV.—Fam. SPARIDÆ or SEA BREAMS.

Body compressed, oblong, covered with scales, the serrature of which is sometimes absent. Mouth and eye like the last. Either cutting teeth in front or molar teeth on the side of the jaws, palate generally without teeth. Dorsal single with nearly equal spinous and soft portions. Three spines on anal. Lower rays of pectoral branched, with one exception (*Haplodactylus*). Ventrals thoracic, with one spine and five rays.

This is a most important family of fishes, which though generally small are nearly all useful. They are divided into four groups, from the form of the teeth. In the first the teeth are incisors in front of the jaws (black-fish of Port Jackson). In the second the teeth are the same, but the pectoral rays are not branched. In the third there is a single series of incisors in front, with several series of rounded molars in the sides. In the fourth the jaws have conical teeth in front, molar teeth on the side, especially adapted for crushing small shells, crabs, &c. This division includes our most abundant market fish, such as black bream, common bream, schnapper, &c.

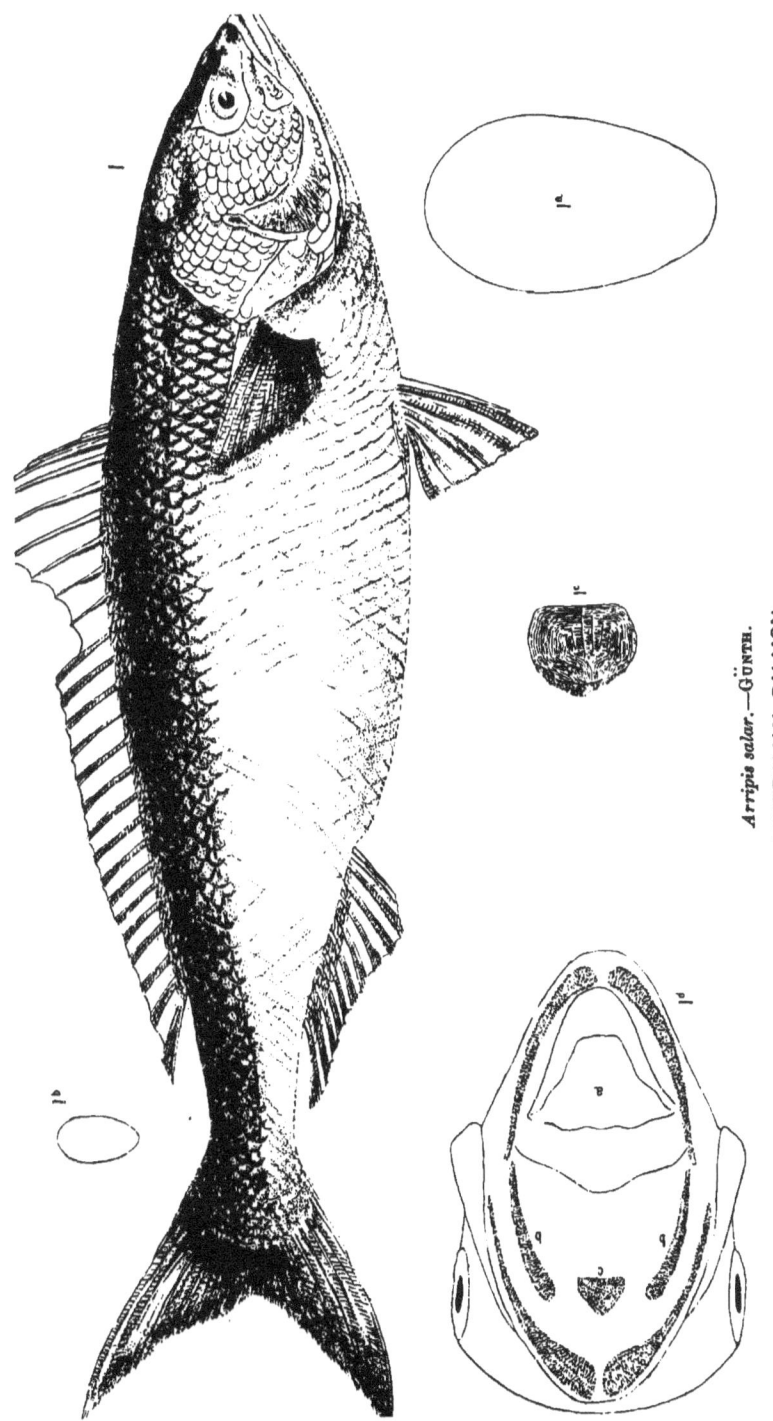

*Arripis salar.*—GÜNTH.
AUSTRALIAN SALMON.

1ᵃ—Section of body.  1ᵇ—Section at caudal end.  1ᶜ—Scale.  1ᵈ—Mouth, showing tongue (*a*) and teeth (*b c*).

## The Black-fish.
(Plate VII.)

The first division includes our Black-fish *Girella tricuspidata*. This genus has scaly cheeks but naked opercles. Dorsal spines received into a rather indistinct groove. Scales moderate. Six branchiostegals. Pyloric appendages numerous, air-bladder divided into two posterior horns. We have two species in Port Jackson, *G. tricuspidata* and *G. simplex*, the main distinction between which is that one has tricuspid incisor teeth and the other has smooth cutting edges with a short series of teeth on the front of the palatine bones. But intermediate forms are found. The black-fish in the market is usually *G. tricuspidata*, but it is quite as common to see *G. simplex* sold as black-fish. In certain seasons they may be caught in abundance in shallow water with the line, the only bait being a green confervoid weed obtained on wood under sea-water. A figure of *G. tricuspidata*, Richardson, is given at Plate IV.

Another poor table fish of this genus is called the "Drummer." It is *G. elevata*, of Macleay. The colour is a uniform brown and the scales are larger. The tricuspid teeth have the middle cusp large and rounded with a few conical teeth on the intermaxillary bone behind. Tail large and the pectoral fins about as large as the head. It is a poor table fish and is caught rarely. Besides the above there are four other species, one of which is a deep blue colour, *G. cyanea*, Macleay. Of the second group, distinguished by the undivided pectoral rays, we have two species, *Haplodactylus lophodon*, Günth. (brown, with a black spot behind, and caudal and anal fins variegated with lighter colour), and *H. obscurus*, Castelnau, a rare fish of nearly black colour.

The fourth group includes all our Schnappers and Breams, and from their interset and utility will require a lengthened notice.

The genus *Pagrus* or as we term it in the vernacular "Schnapper," a word of Dutch origin, is distinquished by an oblong compressed body, with scales of moderate size. Several pairs of strong canine-like teeth in both jaws: molars arranged in two series. Cheeks with scales. The spines of the dorsal fin eleven or twelve in number, though sometimes elongate, and can be received in a groove, anal spines three.

Thirteen species of this genus are known, chiefly distributed in the northern parts of the temperate zone, and more scantily represented in the tropics. Several species occur in the Mediterranean and the neighbouring parts of the Atlantic. One (*P. argyrops*) is well known on the coasts of the United States under the names of "Scup," "Porgy," or "Mishcup." It is one of the most important of food fishes, growing to a length of 18 inches and a weight of 4 lbs.

## The Schnapper, or Snapper.
(Plate VIII.)

The schnapper (*Pagrus unicolor*, Cuv. & Val.) is the most valuable of Australian fishes,—not for its superior excellence, for we have many more delicious, but for the abundant and regular supply which it affords of a very nutritious and wholesome description of food. It is found on all parts of the Australian coast, but most abundantly on that of New South Wales. It is a deep-water fish, found generally on or near rocky points, or reefs running out for miles from the coast. Its food is chiefly the mollusca living on the rocks, though the readiness with which it will snap up bait of the most varied descriptions indicates tastes of rather an omnivorous character. Like all or most fishes, it has its periods of migration and accumulation in shoals, a movement so well expressed by the term "schooling" that we shall adopt the phrase for the future. The time of the appearance of the "school schnapper" is the early part of summer; it is then believed to be at least three years old, the previous stages of its existence being well known under the names of "red bream" at the age of one year, and of "squire" at two. At a still greater age the schnapper seems to cease to school, and becomes what is known as the "native," and "rock native," a solitary and sometimes enormously large fish. The opinion, however, of Mr. M'Carthy, who is undoubtedly a most accurate and intelligent observer, is that these differences actually indicate three or more perfectly distinct species.* At the first appearance of the school fish in early summer the roes are small, but the full size is attained in or about January, about which time no doubt the spawn is deposited. The actual mode of the deposition or attachment of the spawn has never been observed, and the same may be said of the date of the first appearance of the young fry, but there can be little doubt that the deposition takes place in moderately deep water near the land, and that the young are probably hatched before the winter season. The young fish in the shape of "red bream" are abundant in the harbours and inlets, but never in shallow water, and are seldom captured in large quantities in the seine; they take the hook, however, freely, and the capture of them is a very favourite pastime of the Sydney people. The schnapper is generally caught by the hook, but instances have been known of their being taken in Lake Macquarie and other places by means of very deep nets.—R.R.C.

"The schnapper," says Count Castelnau, "is one of the largest and handsomest of the fish of the Melbourne market. It is found all the year round, but those caught in the cold months of the year are generally small; in November and December it becomes much more abundant, and the very large specimens are common. It is a good article of food. As I had already observed at the Cape of Good Hope with respect to *Chrysophrys*, the specimens of this species are subject to very remarkable changes in their form. The female has always a rather oval profile, and the young male has the same, but in the latter sex, age brings on the development of a curious crest on the nape of the head, and a protuberance, which in very old individuals takes the appearance of an enormous nose, and gives to some of these individuals a most remarkable resemblance to the human face. The schnapper is of a beautiful silver pink, with the lower parts of the body white and silvery, dorsal pink, with sometimes white spots on the membranes. The caudal becomes blackish towards its end; all the fins are pink, with the exception of the anal and ventral, which are white. The young specimens are covered with white and sometimes with blue spots, which disappear with age. These appear to be the *Pagrus guttulatus* of Cuvier. Some old specimens assume a beautiful red colour. The species attains large proportions, and sometimes weighs as much as 30 lbs."†

The range of this species is very great. It is found in almost all Australian waters, and extends to New Zealand, Norfolk Island and Lord Howe's Island. Any isolated reef or submerged rock, or as the sailors term it, "bumbora," surrounded by deep water, may be considered its favourite haunt. Subjoined are some very interesting remarks on this

---

* Mr. Oliver denies that there are three species of schnapper; he regards all as varieties, dependent on food and habitat.

† Essay on Ichthyology, in the first volume of the Proceedings of the Victorian Acclimatization Society, p. 70.

SCALE OF INCHES.

*Scorpis æquipinnis.*—RICHARDSON.
"THE SWEEP."

fish and its mode of capture, by Mr. E. S. Hill\* :—"Although by comparison some would imagine that there was more than one of the same genus along our coast, this, however, is not the case; the mere difference of shade in their colour, or lanky or chubby appearance, in all probability is caused by the abundance or scarcity on their feeding-grounds. The schnapper is migratory, and both herbivorous and carnivorous.

"The usual grounds fished in the vicinity of the Heads of Port Jackson are numerous, and for example may be named South Reef, North Head, Blue-fish, Colours, Pine-tree, Tumble-down, and Mud Island, and the wide or Maori ground in deep water, all of which places are known to fishermen by bearing of land-marks.

"The usual method of estimating quantity for sale by the fisherman is, as the schnapper or count-fish, the school-fish, and squire, among which, from its metallic appearance, is the copper-head or copper-colour, and the red bream. Juveniles rank the smallest of the fry, not over an inch or two in length, as the 'cock-schnapper.' The fact, however, is now generally admitted that all these are one and the same genus, merely in different stages of growth.

"The ordinary schnapper, or count fish, implies that all of a certain size are to count as twelve to the dozen,† the shoal or school-fish, eighteen or twenty-four to the dozen; and the squire, thirty or thirty-six to the dozen—the latter just according to their size, the red bream at per bushel. With some exceptions, these are usually caught in the vicinity of reefs or rocks, by anchoring your boat according to the turn of tide, so that your line will trend to or along the rocks. Occasionally, when migratory, they are caught further off in deep water, but when in large shoals, wide off and on the top of the water, which is termed schooling, they will rarely take the bait.

"Apart from these, however, the largest of the genus is more frequently caught off points within our harbours after nightfall, and at certain time of tide—young flood at some, ebb at others. A fine old fish, monarch of his ground, with a figure-head as bold and defined in its outline as that on the North Head of this port, cunning and fastidious to a degree, a regular epicure—he must be tempted with choice morsels before the well-chosen bit which conceals the hook will engage his attention. And you must bear in mind at the same time that (as old Charley Reynolds used to describe him) he is only a oncer, implying that if he once fairly took the bait, and you pricked or missed him in the strike, it was good-bye for that time; but fairly hook him with strong tackle your work and anxiety for the time it lasts will be somewhat intense. The tugging, jerking motion of the schnapper is unmistakeable, and when he gets his shoulder to the line he goes off with a rapidity that makes the cord whistle again, either through your fingers or over the boat's gunwale; a steady and continuous strain, no stray line, together with some skill, enables you to safely land him, at which time you can realize that your patience, toil, and anxiety, are rewarded with a fish from twenty to twenty-five pounds weight, fit to embellish a noble banquet.

\* Fishes of and Fishing in N. S. Wales. *Sydney Mail*, 1874.
† Count fish are usually 6 lbs. each, so that a dozen equals 72 lbs.

"The bait for these fishes are star-fish, squid, mackerel, yellow-tail, mullet, tailors, and a variety of other fishes; the whole of these at particular times will be readily taken, but when the schnapper appears dainty, mackerel and squid may tempt him when all others fail. In their young stage, such as red bream, lean beef, for amateurs, is about the usual bait, and which answers in the absence of fish tolerably well. Thousands of what would be the future schnapper are destroyed in this harbour alone, either by the line or net, or by the combination of both, from the very smallest of the genus, and in their earliest stage they appear willing to take the bait.

"It may be fairly presumed that this fish is known to almost everyone in the Colony, and often recognized in the adult state, and when just caught is the beau ideal of a fish, representing beauty to the eyes, sport to the fisherman, firm, palatable, wholesome, and nutritious food to the multitudes. Cook it whichever way you may please—by the primitive and impromptu method frequently adopted by fishermen, of roasting it before a fire, elevated an inch or so from the ground by the aid of a couple of forked sticks, which answer the purpose also of turning the fish by lifting them from the ground and reversing their position. No preparation is necessary in the first instance, save to take out the inside and wash the part clean; when sufficiently cooked, the scales will come off from either side in one flake, leaving a firm, beautifully white and tempting dish. Then for the table, boiled entire, and served with egg sauce, in fillets or as a curry, all are excellent, and as a general rule the schnapper may be ranked as one of the first favourites. No party of amateurs who go out for a day's general fishing think their basket complete without schnapper being among their number,—that name floats uppermost in calling over the day's sport, and covers many defects in other fishes which may happen to be in the list of the day's catch, and indicates, as a rule, that the sport was fair or good."

In the frontispiece is given a figure of an old male schnapper, with the peculiar hump on the head. It is formed by an egg-shaped mass of bone at the summit of the high crest of the supra-occipital. The frontals and ethmoid also become enormously thickened, and there is frequently a large egg-shaped mass on the end of the first interneural spine. The aboriginal name of this fine fish is "Wollomai."

## The Tarwhine.

### (Plate IX.)

CHRYSOPHRYS comprises the Tarwhine and Black Bream of the Sydney fishermen.—This genus only differs from *Pagrus* by the upper molar teeth, which are in at least three series, while in true schnapper they are only two. The air-bladder is notched or has small appendages in bream, whereas the other genus has the same organ quite simple. Some twenty species are known in tropical seas and the warmer parts of the temperate zone. The common species of the Mediterranean is *C. aurata*, rarely found on the south coast of England, where it goes by the name of Gilthead. It was known to the Greeks by the name of *Chrysophrys* (golden eyebrow), and the Romans, as *Aurata*, or golden, and the French name *Daurade* appears to be a translation of the same. According to Columella the *Aurata* was one of the fishes reared by the Romans in

PLATE VII.

their Vivaria, and the inventor of these Vivaria was named *Orata*, from this fish. It grows extremely fat in artificial ponds. Duhamel states that it stirs up the sand with the tail so to get at the shell fish. It is extremely fond of mussels, and often makes its presence known by the noise caused in breaking in the shells with its broad teeth. (G.S.F.) These facts may give a clue to the habits of our own species.

We have two species in Australia, as already noticed. The Black Bream, *C. australis*, Gunth., and the Tarwhine. *C. sarba*, Forsk., *C. hasta*, Bleek., is a common fish in India and China, entering all the large rivers. It was supposed by Mr. Hill to be our common species, but Günther has shown that they are distinct. The Australian Bream, Plate VI, is as common on the south as on the east coast. It affords excellent sport to anglers in Victoria. The author remembers in January, 1860, catching an immense quantity with a line in the Glenelg River, Victoria, where the river was little more than brackish, though not far from the mouth. The bait used was a small crab, and no sooner was the line down than the hook was swallowed. The sport was only terminated when all the hooks were destroyed, for these ravenous fishes broke them with their hard teeth. As a food-fish it has good qualities, and is very abundant, but never more than about a foot in length.

It is somewhat difficult to distinguish between the two species, so the definitions of both are given. *C. sarba* has the height of the body nearly two-thirds of the length, and the head is one-fourth. The distance between the eyes is rather more than their diameter. Incisors broad, obtuse, the molars in great number, forming four series in both jaws, and a large ovate molar behind. Dorsal spine, moderate, rather compressed, broader on one side, the fourth longest, second and third anal spines nearly equal. Silvery in colour, with about fourteen indistinct longitudinal streaks.

*C. australis* has the body proportionately slightly higher, and the eye a little longer. The dorsal spines are stronger. The second anal spine is very strong, and equal to the fourth dorsal in length. There are five series of scales between the preorbital and the angle of the operculum. Colour, silvery and grey on the upper parts. There is a slight brown transverse band on the forehead ; dorsal, hyaline, bordered with black ; caudal rather yellow, with a dark external border ; anal either yellow or dark ; ventrals yellow, or in part blue. Whole body with shining longitudinal streaks, but this character is often absent from the Victorian specimens.

The "black bream" (*Chrysophrys australis*) and the "tarwhine" (*Chrysophrys hasta*) are both excellent fishes, and are frequently abundant. The schooling season seems to be summer with both species, but where or when they spawn has not been ascertained. They have been occasionally caught outside on the schnapper grounds, but their chief resort is evidently the harbours and lakes along the coast, where they are taken in the seine in great numbers. In Port Jackson line-fishing for "black bream" is a very favourite sport.—R.R.C.

## The Silver Bream or Silver-belly.

Mr. Hill, in the series of essays already referred to, speaks of a silver bream or white bream. It is probable he refers to *Gerres ovatus*, a common fish of very compressed form, and very protractile mouth.

They probably never enter fresh water. There are about thirty species known, all of which have a plain silvery colour, with smooth or ciliated scales, eye large, dentition feeble, and palate toothless, dorsal fin nearly divided.

The white bream, the subject under notice, is scarcely so well known as either the red or black bream, neither is it so great a favourite as an article of food. This fish is seldom or never caught by hook and line. The net is the only certain means to bring a supply, and then at particular seasons only. It is necessary to cook the silver-belly, as it is often called, perfectly fresh, and the sooner after it is caught the better; otherwise, if allowed to get stale, it is flavourless, flabby, and soft, and more like some of the muddy fresh-water fishes in taste. In this species the lower pharyngeal bones are coalescent, which makes its place among the Perches doubtful.

The same author gives an admirable sketch of the habits and mode of capture of the black bream. He says that these fishes visit our harbours "from seaward periodically during the summer months of February, March, and April, and are adult fishes, being full-roed at the time of their visit; probably they are migratory for the purpose of spawning. These are fine conditioned, and firm, good-eating fishes, and ranked by a majority as one of the best. During the period of their visit they are terribly harassed by hook and line and the net. Scores of boats wait on them during the night at Watson's Bay, Camp Cove, Quarantine, and Middle Harbour Points, armed with good and very light tackle—the finer the better; and for bait, some fresh or a day old mackerel, or much better still, some prawn procured at low-tide by stamping down the weeds to muddy the water; these are called nippers, from the disproportionate size of one claw. The black bream likes a large soft bait as a rule, but on these occasions he seems to prefer the crustacean named; they pick it up and rush off at full tilt (no nibbling with them), and pull hard and sheer about with a full determination if possible to get away. Some difficulty is experienced by the novice in attempting to unhook his fish, and often a lacerated hand is the penalty of his want of skill. The black bream is armed with spines, as the order implies; but his are sharper and more robust than any of the others, and the first rays of the pectoral and ventral fins were used by the aboriginals wherewith to tip their fish spears. To unhook a black bream it should be firmly grasped about the middle—certainly about the pectoral fins, sufficient to paralyse him, then take out the hook. Otherwise, if the fish is handled gingerly, the first wriggle he gives will release him, and the second motion will perhaps cause your hand to be cut deeply by the operculum or bony gill covering, which is as sharp as a good knife, or stab you with a spine or two in the struggle. The black bream, however, which remain and are plentiful in the harbour at most times, become terribly gross and filthy feeders. Nothing seems to come amiss to them, and they are found more plentiful about the mouths of sewers, hulks, and ships, which remain stationary for any period, under wharfs, and near bathing-houses. They are often sought at these places by the amateur. These fishes exhibit a great deal of cunning, and require an artificial bait to tempt them. Dough amalgamated with herring, anchovy, or ling. With this the hook, already

PLATE VIII.

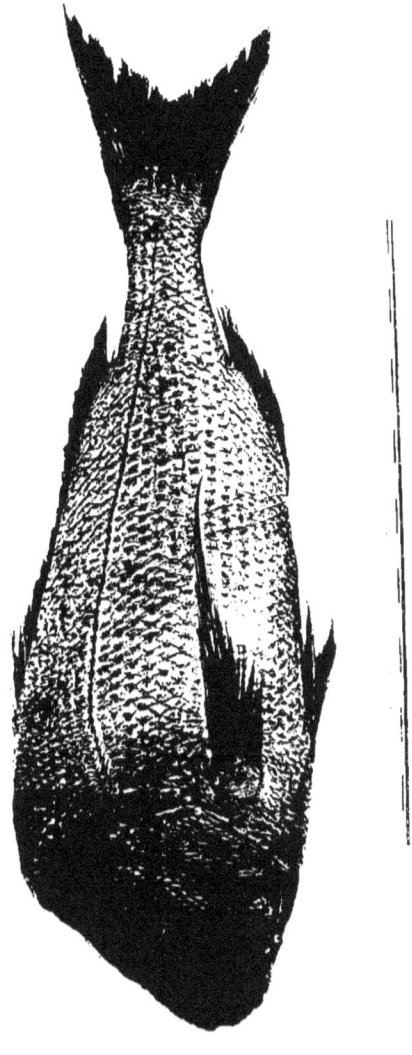

*Pagrus unicolor.*—CUV. & VAL.
THE SCHNAPPER.

FISH AND FISHERIES.            45

snoozed with gut, is concealed and carefully lowered down into the water. The bait, like a plump oyster, rarely fails to attract; but very often the darkie (as the bream is familarly called) seizes the bait and comes off best. Dark nights and flood tide afford the best time for black bream fishing. At such times it is usual to have a lamp in the bottom of the boat to enable you to disentangle your line or examine your bait, &c. This, however, is not indispensable; but in its absence it certainly is advisable to have a set of lines ready for use, so that when one is disabled it may be bunched up and another used. Black bream fishing, when they bite freely, is first-rate sport, and which is much enhanced by having one party in the boat for the express purpose of making ready your bait and attending to the lines. Notwithstanding the cunning and shyness of these fish during the day, they are readily caught at night by baited baskets or traps lowered to the bottom. A junk of beef boiled almost to tatters is secured inside to the bottom of the trap. The small fishes are first attracted. They enter and tug away at the bait, which is easily shredded. The bream soon follow and are captured. Meshing nets, placed along the rocks, also secure many at night, but the ordinary hauling net or seine very frequently secures great numbers.

The Bream is found in the Hunter River even where the water is nearly fresh but most abundantly where it is brackish. The anglers in that locality find that the best bait is a prawn (*Penæus*), but it must be boiled or the fish will not touch it.

## V.—Fam. HOPLOGNATHIDÆ.

Only one species in Australia, of no economical importance.

## VI.—Fam. CIRRHITIDÆ.

Body oblong, compressed, with cycloid scales, lateral line continuous. Mouth and eyes as the last families. No bony stay for the pre-operculum. Six (generally), five, or three branchiostegals. Dentition more or less complete, composed of small pointed teeth, sometimes with the addition of canines. Dorsal fin single, of equal spinous and soft portions. Anal, with three spines, generally less developed than the soft dorsal. The lower rays of the pectoral fins simple, and generally enlarged; ventrals thoracic, but remote from the rest of the pectorals, with one spine and five rays.

The fishes of this family may be readily recognized (says Günther) by their thickened undivided pectoral rays, which in some are evidently auxiliary organs of locomotion, in others probably organs of touch. They differ from the following family, the *Scorpænidæ*, by the absence of a bony connection between the infraorbital ring and the pre-operculum.

This is a small but natural family, and is well represented in Australian waters in the genera *Latris* and *Chilodactylus*. The first of these is the genus of the well-known "Hobart Town Trumpeter," a fish deservedly of high reputation, and of three other species added by Count Castelnau to the Victorian fauna. The other genus (*Chilodactylus*) is also largely represented in Tasmania and Victoria, one species being commonly imported from Hobart Town in a smoked and dried state under the name of "perch." The species found in New South Wales are, two fishes called by the fishermen "morwong," both of the genus *Chilodactylus*. They are very good fishes in the fresh state, and equal to the cod or ling for salting, but unfortunately they are rare. They are generally caught in deep water on the schnapper grounds, and it is quite probable that they may be found more abundant towards the southern limit of our sea-board than they are here, but our inquiries have not been successful in eliciting information in that direction. Another species of *Chilodactylus*, known among the craft as the "carp" is more of a rock-fish, being frequently taken in the harbour in nets, but it seems to be only an occasional visitor. The only species of *Latris* (*L. ciliata*) found on this coast is extremely rare, and certainly not to be compared for a moment to its Tasmanian relative *Latris hecateia*, the "trumpeter."—R.R.C.

The two genera thus referred to are thus described :—*Chilodactylus*. One dorsal fin, with from sixteen to nineteen spines; anal fin of moderate length, caudal forked. One of the simple pectoral rays more or less prolonged, and projecting beyond the margin of the fin. Teeth in villiform bands, no canines. Pre-operculum not serrated. Scales of moderate size. Air-bladder with many lobes.

Seventeen species are known, chiefly from the temperate parts of the South Pacific and the coasts of China and Japan. They belong to the most valuable food fishes, as they grow to a considerable size, from 5 to 25 lbs., and are easily caught in numbers. At the Cape of Good Hope they are very abundant, and preserved in large quantities for export.

## The Morwong.

(Plate X.)

The Carp or Morwong (*Chilodactylus macropterus*, Richardson) has six simple pectoral rays, the uppermost very long; dorsal fin notched, the seventh spine longest, higher than the soft dorsal, and one-half the length of the head. The second anal strong and longer than the third. A blackish band from the origin of the dorsal to the base of the pectoral. Length, 12 inches, rarely 18.

Count Castelnau says that this fish is called "Bastard Trumpeter" by the Melbourne fishermen (but the name is also applied to *Latris fosteri*); and adds that the colour is silvery, with the upper parts and head of a light purple; a black spot behind the upper part of the operculum; branchiostegal membrane beautiful light blue; dorsal, caudal, and anal fins of rather dirty yellow, with the spines purple, ventrals white, pectorals yellow, with white interior rays; eye silvery, with a blue ring. Some iridescent longitudinal streaks on the sides. In this Colony the Morwong is also called "Jackass-fish."

## The Red Morwong.

(Plate XI.)

The red Morwong or Carp, *C. fuscus*, Castelnau (plate VIII), is of a uniform reddish colour, and about the size of the last species. There are six simple rays in the pectoral fins, the upper one not much longer than the branched ones, the following very long, the others gradually shorter; dorsal scarcely notched, spiny portion seventeen spines, first rather short, fourth longest, back gibbous, ridge on front edge of orbit, caudal strongly emarginate, anal with three spines and nine rays; the eye is

very prominent. This is one of the most delicious fish of our seas; it is often caught in nets set near the headlands of Port Jackson, as it is peculiarly a rock-fish. The aboriginal name of this fish is "Bingátti."

*Chilodactylus vittatus*, Garrett, or Banded Morwong, Plate XIII, is another kind which is rarely seen in the markets.

## The Trumpeter.

LATRIS (Trumpeter) has the dorsal fin deeply notched, the spines seventeen; anal, many rayed, none of the pectoral rays extending beyond the margin of the fin. Teeth viliform, no canines. Preoperculum minutely serrated.

These are considered amongst the most delicate of our edible fishes, but one species, *L. ciliaris*, is not much valued. It has seventeen spinous and thirty-nine soft rays in the dorsal fin, and three spinous and thirty-two soft in the anal. Head only one-fifth of the whole length.

## SCORPÆNIDÆ.

Body oblong, more or less compressed, covered with ordinary scales or naked. Cleft of the mouth lateral or subvertical. Dentition feeble, consisting of viliform teeth, and generally without canines. Some bones of the head armed, especially the angle of the preoperculum; its armature receiving the support of a bony stay, connecting it with the infra-orbital ring. The spinous portion of dorsal equally or more developed than the soft and than the anal. Ventrals thoracic, generally with one spine and five soft rays or rudimentary.

This family consists of carnivorous marine fishes only; some resemble sea perches in form and habits, whilst others live at the bottom of the sea and possess in various degrees of development those skinny appendages resembling the fronds of seaweeds, by which they either attract other fishes or hide themselves. These species also resemble in colour their surroundings, and vary with change of locality. They move or feel by means of simple pectoral rays. Nearly all are distinguished by a powerful armature, either of the head or fin spines, and in some the spines are developed into poisonous organs. (G.S.F.)

Under the head of TRIGLIDÆ or Gurnards were formerly included the above family. Now we have only one or two of our edible fishes included in it, that is to say, the genera *Sebastes* and *Scorpæna*, both of which comprise the red rock cod of our fishermen. They are rock and ground fish, but soft, and seldom eaten.

SEBASTES has the head without a groove on the occiput and few small spines. We have only one species, *S. percoides*, Richardson, which is a reddish fish with five brown cross bands.

## The Red Rock Cod.

(Plate XII.)

This fish belongs to the genus *Scorpæna*. It has a naked groove on the occiput, armed with spines and sometimes skinny tentacles. About forty species are known from tropical and sub-tropical seas. They lead a sedentary life, lying hidden in the sand or between rocks covered with seaweed, watching for their prey, which chiefly consists of small fishes. Their strong undivided pectoral rays aid them in burrowing in the sand and moving along the bottom. Their colour is very much the same in all species, that is an irregular mottling of red, yellow, brown, and black, but the pattern varies exceedingly in the same species, and even in the same individuals. They never exceed a length of 18 inches, and even that size is rare. The flesh is much liked by some. Wounds inflicted by their fin spines are exceedingly painful, but not followed by serious consequences. (G.S.F.)

We have four species of this genus in Port Jackson. Plate IX represents *S. cruenta*, Richardson. The colour of this species is a beautiful scarlet, sometimes marbled with grey ; belly, whitish; sides, with rounded dark blotches ; fins, reddish pink variegated with white ; the spinous dorsal has a large black blotch covering the upper half of the posterior portion, the soft dorsal is thinly spotted, and transversely marked with white and red ; ventrals, pink ; pectorals, marbled with pink, white, and brown.

At Plate XIV a figure is given of *Sebastes percoides*, a fish of a closely allied genus of the same family. It is caught at times in Port Jackson, but has no local name. In Victoria it is called the Red Gurnet Perch.

## The Bull-rout.

This is the name given by the fishermen of the Hunter River to a fish belonging to the family SCORPÆNIDÆ, known to naturalists as *Centropogon robustu*. It is a small fish, seldom attaining 7 inches in length or exceeding 5 or 6 ounces in weight. Like all the scorpion-fish it is very ugly, with prominent gaunt ghost-like eyes set in large hollow sockets. Its colour also is dull and dirty-looking, as like as possible to the brown, green, and black slimy weeds in which it hides with security. There are two remarkable peculiarities about this fish. One is that it emits a loud and harsh grunting noise when it is caught, so that if by chance it takes a bait, the fisherman knows what he has got by the noise before he brings his fish to the surface of the water. When out of the water the noise of the Bull-rout is loudest, and it spreads its gills and fins a little, so as to appear very formidable. Another peculiarity is that the spines about the head are venomous, and inflict most painful stings. If a Bull-rout is taken (when dead of course for safety sake) and held up to the light with its back towards you, a long curved stout spine will be noticed on each side of the snout just above the mouth. There is also a little spine on each side in front of this. Then on the preoperculum there is a row of four spines, increasing in size as they go backwards. Now if any of these spines chance to wound you, which they may easily do, for they are as sharp as needles and very strong, the pain is intense. It runs through the whole limb like fire. The injured part becomes red and inflamed. But except the pain, which all victims assert is very agonizing, there are no serious occurrences. There is also a constant testimony that the pain usually ceases at sunset, and however strange it is, the fact is confidently asserted by those who have had experience. The Bull-rout is never caught, except accidentally, as

*Chilodactylus macropterus.*—RICHARDSON.

CARP OR MORWONG.

it is quite valueless. It is very often brought in when fisherman are trying to net prawns for bait. It will take a small prawn bait itself, and thus becomes a great annoyance to those who are angling for Perch (*Lates*), as it is very numerous. Its sting is most frequently felt by bathers, who tread upon it as it lies on the bottom amongst the weeds. The blacks held it in great dread, and the name of Bull-rout may possibly be a corruption of some native word. The venom is probably a mucus secreted by the skin, and not connected with any distinct poison gland.

In Port Jackson there are other species of the family, and other *Centropogons* which need not be enumerated here. *Pentaroge* is, however, a genus which merits a passing notice. It has no scales and has strong spines on the preoperculum and preorbital dorsal fin, with twelve or thirteen spines, anal with three. No pectoral appendages; viliform teeth on jaws, vomer and palatine; air-bladder small; pyloric appendages in moderate number; a cleft behind the fourth gill.

## The Fortescue.

The "Fortescue" (*P. marmorata*, Cuv. and Val.) is a common fish in Port Jackson, with a very long dagger-shaped preorbital spine, cleft of the mouth slightly oblique, and a dorsal fin which begins on the neck. The colour is a dirty yellowish, largely marbled with brown. The author when fishing once saw a man wounded in the hand by the spine of this fish, and for an hour or two suffered intense pain, but no worse consequences followed. Of this fish, Mr. Hill says: "The scorpion or Fortescue, as these fish are popularly termed by fishermen, have been known for a long time, and bear that name no doubt in memory of the pain they have hitherto inflicted; and for its number and array of prickles it enjoys in this country the *alias* 'Forty skewer' or 'Fortescure.' These are net fishes, and the moment they are captured they set up their thorns and remain rigid, only waiting for some one to handle them.

The thorns with which it is armed cover nearly all the fish when in this state, and one is exposed at every touch. These are small fishes, and weigh but a few ounces, and at night fishing they often get rolled up in the net which when being hauled in bears some pulling to get it ashore. They are easily distinguished from *Centropogon* as they have no scales, but in other respects the Fortescue is very like the Bull-rout.

"We were out fishing one night with a net towards Chowder, a bay in this harbour, and were accompanied by a blackfellow named Wallace. He got hold of one of these fishes, which was in the net rolled up, and he had put his whole force and pressure upon that spot. I never saw any one in such pain for a short time. He rolled on the beach, then got up and ran about like mad. I was necessitated to give him at intervals all the grog that we had, and which consisted of nearly a quart of strong spirits. This was scarce enough to cause the pain to leave, but it had the effect of deadening it, and in a couple of hours we were enabled to remove him to his camp, when a good sleep and the effects of the grog put him right again. Strong ammonia is the best thing which can be applied to these parts when stung with a fish-bone."

G

Two families follow in this division, and then we pass to the tenth and last of the perch-like Acanthopterygians,

## The TEUTHIDIDÆ.

Body, oblong; strongly compressed, covered with very small scales, lateral line continuous; a single series of cutting incisors in each jaw; palate, without teeth; dorsal, single, the spinous portion most developed; ventral fins thoracic with an outer and an inner spine with three soft rays between.

This family consists of one very natural genus, *Teuthis*, easily recognized by the singular structure of the fins. The incisors are small, narrow, and with a serrated edge. The air-bladder is large, forked before and behind. The skeleton is very peculiar, with twenty-three vertebræ, ten of which belong to the abdominal portion. There are thirty species known, all herbivorous, and do not exceed 15 inches in length.

## The Black Trevally.

The "Black Trevally" of the Sydney fishermen is *T. nebulosa*, Quoy and Gaimard, a brown fish, irregularly marbled. It extends right round to the tropics. Of this species Mr. Hill says:—"These are net fishes, and are often caught by that means in large numbers, but are not good market fish, and soon after being caught they look dirty—consequently they are not esteemed as good fishes."

"The black trevally is shaped something like the white one, but it is slimy and more rigid, having all its prickles pointed, and has also very small scales. These fishes appear very different, and the dorsal ray, as well as every fin, is not only kept stiff, but each prickle or point of any one inflicts a very painful wound when it pierces.

"I recollect many instances of only slight touches, but the pain lasted long; and on one occasion we were net-fishing in North Harbour, a favourite resort for these fishes, and a goodly number was hauled in. A young gentleman who had accompanied his father on the expedition was cautioned at the time against handling any of these fishes; he however made a kick at one, and a spine went through his boot and sock, and penetrated the ball of his toe. In less than a minute the boot and sock were off (from pain), the part was well washed with spirits, but to no effect. The father had to put the toe in his mouth and well suck the punctured part, and continually to apply spirits, and not for half-an-hour did the pain commence to subside.

"The black trevally is a very good-eating fish when used at once. When alive all the fins are cut off with a pair of large scissors we usually provide for that purpose. When the fish is cleaned well, and fried in good clear dripping or butter, adding pepper and salt to the palate, you will find it really good, but they must be used fresh to get out all their nice qualities."

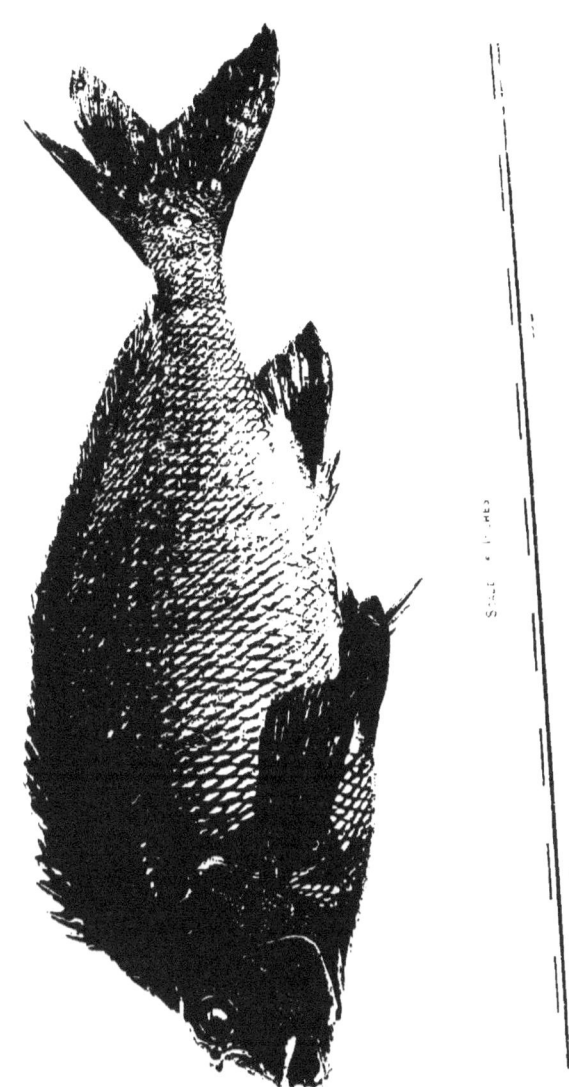

*Chilodactylus fuscus.*—CASTELNAU.
CARP OR RED MORWONG.

FISH AND FISHERIES.  51

## The White Trevally.

This is another species, the "White Trevally,"* or *Teuthis javus*, Linn. The ground colour is dark with whitish spots, round on the back, confluent at the side, and forming large natural streaks on the belly—spots and streaks narrower than the interspaces, vertical fins without spots.

## II.—Div. ACANTHOPTERYGIANS.
## Fam. BERYCIDÆ.

Body short, with ctenoid scales, rarely absent. Eyes lateral, large (except *Melamphäes*). Cleft of mouth lateral, oblique, jaws with viliform teeth, palate generally toothed. Opercular bones more or less armed. Eight branchiostegals. Head with large muciferous cavities covered with thin skin.

Marine and generally deep-water fishes, frequenting warmer temperatures, and having a wide range, some species being common to Madeira and Japan. This family is most abundant in tropical seas, three genera only being represented in New South Wales waters—*Trachichthys, Beryx*, and *Holocentrum*. Of these one only merits notice—it is the well-known "nannygai," *Beryx affinis* of Günther. It is a deep-sea fish, caught only with hook and line, and rarely taken during the winter months. It seems generally to make its appearance soon after the commencement of the warm season, and to judge by the number sometimes taken at one time, probably comes in considerable shoals, but we have not been able to determine whether it is a migratory fish in the true sense of the term, visiting these temperate seas in the summer and returning in winter to warmer latitudes, or whether its appearance is merely, as is the case with very many fishes, a movement only from deep water to the neighbourhood of the land. The same uncertainty exists as to its spawning season. It is seldom seen full-roed, and we have not found that the very young fish are ever seen in our harbours or on our coasts, so that it may be fairly inferred that the "nannygai" does not as a rule spawn in this vicinity. We have no evidence of its having been found south of Jervis Bay, and we know that it is unknown in Port Phillip, but to the north its range seems to be unlimited. As an edible fish it ranks high, indeed there are few better in the country. It cannot however be much depended upon for the market, owing to the irregularity of the supply, but this may proceed not so much from the scarcity of the fish as from the fact that the fishermen never specially seek it, and that it is only caught accidentally when fishing for schnapper. The "nannygai" when slightly corned and smoked is said to be a great delicacy.—R.R.C.

The genus thus referred to has a very short and obtuse snout, prominent chin and large eye, viliform palatine and vomerine teeth; a strong spine at the scapulary and angle of preoperculum, rather small scales, abdomen serrated. Dorsal with from three to six spines, ventral with six soft rays. Four species known from New Zealand to Madeira.

## The Nannygai.
(Plate XV.)

*Beryx affinis*. Günth. Catal. The "Nannygai." Dorsal, seven spines and twelve rays, first highest, caudal very deeply forked, of four spines and twelve rays. Ventrals with one strong spine and seven rays, pectorals with thirteen rays. Colour, a most beautiful pink, with silver stripes on the body and edges of the opercula. Length, from one to two feet. Count Castelnau doubted if this Sydney fish was not

\* So called by Mr. Hill.

different from Günther's *B. affinis*, because the lobes of the tail were often so unequal, the lowest being shorter, but this is subject to great variation in the species.

The name of Nannygai is said by Mr. Edward Hill to be derived from the aboriginal name of *Mura nyin a gai*, whatever that may mean. Amongst the early colonists it used also to be called "Mother nan a di" probably a corruption of the native name. The following remarks of the same author will be read with interest :—

"That which is taken on our coast visits some of the headlands once or twice during the year, is shaped like a squire or small schnapper, and does not attain a great size; its colour is bright red, with iridescent streaks, and the colouring matter appears to be in the skin and epidermis, as they retain a portion even after having been cooked.

"The nannygai has a very large eye, and is found in the vicinity of reefs, no doubt for protection, and is first fished for in deep water at the North Head of Sydney, in the month of October, with hook and line, using the ordinary schnapper bait. Great caution is necessary at times in approaching the ground so that the boat may not be pulled over the rocks where they congregate, as these fish have been known to take alarm in such instances, and in consequence not to take a bait during the day; otherwise they bite freely, and many are taken. A month or so later a few are found off Middle Head, and some have been caught between Shark Island and Milk Beach. The North Head of Botany Bay also affords a rendezvous for these fishes, as also at intervals along the coast, south.

"The beauty of this fish is to be seen only when first caught, and the flesh is much better when fried very fresh; in fact, most of the Australian fishes are tenfold better as food when just caught than when stale, although some keep tolerably well; but it cannot be expected that the nannygai, which puts in an appearance in the month of October, can keep many hours after removal from its element."

The division of Kurtus-like Acanthopterygians may be passed over. The division which contains the family Polynemidae is for Acanthopterygians with two rather short dorsal fins, somewhat remote from each other; free filaments at the humeral arch, below the pectoral fins; muciferous canals of the head well developed. Tropical fish in brackish and fresh waters or very muddy rivers, the eyes are often covered with a filmy skin, but the long filaments seem to act as feelers instead, where eyes would be of little use. The flesh is esteemed, and the air-bladder is valuable. They sometimes obtain the length of 4 feet.—(G.S.T.) An instance has been mentioned in the introduction to this essay of a purblind fish of this family having been found at Port Darwin.

Three species of *Polynemus* have been seen on this coast, but their appearance is extremely rare, and as useful fishes they are unknown. We mention them however because the group of fishes to which they belong are of the greatest service in the tropical seas of India and Polynesia, some of the species yielding food of the most excellent quality, and a few of them having air-bladders of good size and average quality for isinglass. It is not improbable that some of the species may be found to be abundant on the northern coasts of Queensland.—R.R.C.

Though three species are here referred to, only two, according to Macleay, have been found—*P. indicus* (Shaw) and *P. macrochir* Günther).

*Scorpaena cardinalis.*—Richardson.
RED ROCK-COD.

In the next few divisions of the Acanthopterygians there is only one family of interest to the Colony, so this need alone be referred to.

## SCIÆNIDÆ.

Soft dorsal more developed than spinous or anal. No filaments, head with well developed muciferous canals ; scales ctenoid. Lateral line well developed and frequently extending over the caudal fin. Eye lateral, of moderate size. Teeth in villiform bands, no molars or incisors, canines sometimes present, palate toothless, preoperculum unarmed. Stomach cœcal. Air-bladder often with numerouns appendages. Coast fishes near rivers or exclusively inhabiting them, tropical or sub-tropical, with very wide specific range, sometimes very large ; all edible. Dr. Günther says they are rare in Australia, but this is hardly correct. Two of our largest market fishes belong to the family, the "Jew-fish" and the "Teraglin."

## The Jew-fish.

(Plate XVI.)

This fish is of the genus *Sciæna*, which has either an overlapping upper jaw or both jaws equal. Interorbital space moderately broad and slightly convex. Outer series of teeth larger than the rest, but no canines, no barbels. Some fifty species are known, among which is the "Meagre" or "Maigre," attaining the length of 6 feet. It has been found in European seas, at the Cape, and in South Australia—at least this is Dr. Günther's opinion, who unites our *Sciæna antarctica* (Castelnau) with the Meagre (*S. aquila*). Count Castelnau, however, points out the difference. He says on comparing the fish with the descriptions of Cuvier and Günther, he even doubts if it belong to the same genus. The maxillary does not reach the eye, and the eye is proportionately smaller. The colour also different ; the back is blue, changing to green, the sides and lower parts of a dirty white, rather silvery, dorsal, anal, and ventral fins reddish, pectorals whitish, with their extremities dark ; some sinuosities or notches at the angle of the pre-operculum.

The jew-fish is a fish that attains a great size, even to 5 feet in length, and always finds a ready sale in the market. It is said to be very frequently cooked by fishmongers and others and sold as "fried schnapper." It is found at almost all seasons, but most abundantly in summer. It is a deep-sea fish and caught only by the hook, though young specimens may be occasionally taken in the net inshore. Of its history and habits nothing further is known. It is found in Victoria, but not in abundance. It is the "king-fish" of the Melbourne market. It has been more than once asserted that this fish is identical with *Sciæna aquila*, the well known and highly appreciated "maigre" of the Mediterranean, and Count Castelnau, though originally describing our jew-fish as a different species, has lately admitted having some doubt on the subject.—R.R.C.

In alluding to the ordinary jew-fish, Mr. Edward S. Hill says "that it attains an immense size, and is caught in our bays and harbours. Care must be taken in a description that I do not cause it either to be confounded with the teraglin or the silver jew, as there is a wonderful difference in the value of these fishes in a gastronomic point of view, an ounce of the latter being preferable to a pound of the former. An old jew-fish, and what might be called a large one, will weigh 50 to 60 pounds; the scales are very large, the teeth formidable, flesh coarse, and sometimes filled with parasites in the form of worms. This fish is not looked upon with any degree of regard, and is rarely sought after, except for an exhibition of his great strength, and which in my estimation is about the only redeeming quality he has. A true carnivora, formidable to shoals of smaller fishes on which he attends for his meals, making havoc among them as occasion and his good appetite may require, these adult fishes exhibit a good deal of cunning, and prefer turbid waters for their depredations; they will go high up rivers, even into fresh water, and make an astonishing commotion among the shallows. The half-grown ones frequently hunt in numbers like a pack of wolves, or our own native dogs bailing up an old man kangaroo in a waterhole. I recollect on one occasion, in the basin at Broken Bay, when some half dozen of us were on a week's fishing excursion, armed *cap-a-pie* with boats, nets, lines, spears, and all the necessary paraphernalia for the venture, and anchored a short distance only from the entrance. At the dawn of day, and young flood-tide, we saw a terrible commotion among the smaller fishes; the net-boat was immediately equipped, and we at once pushed off for the entrance or channel, which leads into the basin. The tide, which had risen about a foot on the flats inside, afforded shelter for the time for the smaller fishes, but as soon as the depth increased a few more inches, in went the pack of jew-fishes pell mell, which created a sensation among, and actually drove many of the fry high and dry ashore. Our net was speedily run out, and it was sufficiently long to cut off the retreat of four or five and twenty of the medium-size jews; their little game was at an end, and after several desperate attempts to get out we succeeded in securing the lot, besides a large number of various fishes which they had hemmed in.

"The deep water of Middle Harbour, above the Spit, and along the rocks high up near the Echo Point, are favourite resorts for these huge fishes, and it requires good lines and hooks to resist the heavy drag they put on.

"Sydney harbour is a great place, off the points, where the water is deep and the tide strong. Here they lie in wait for their prey, but after rain, when the water is muddy or discoloured, they sally out from these lurking-places in every direction. Early morning or nightfall is the favourite time for fishing for the jew-fish."

The "Teraglin" belongs to the genus *Otolithus*, of which about twenty species are known from the tropical and subtropical parts of the Atlantic and Indian Oceans. The air-bladder is of extraordinary form, with appendages above and below. Snout obtuse or somewhat pointed; lower jaw longer. First dorsal with nine or ten feeble spines, canine teeth; preoperculum denticulated; scales moderate or small.

## The Teraglin.
### (Plate XVII.)

*O. otelodus*, Günth. Catal. (*O. teraglin*, Macleay), is a bluish silvery fish, lighter on the belly. All the fins of a dark tinge except the ventrals, which are white, with the space between the first and second rays black. Teeth acute, recurved; eye large, preoperculum

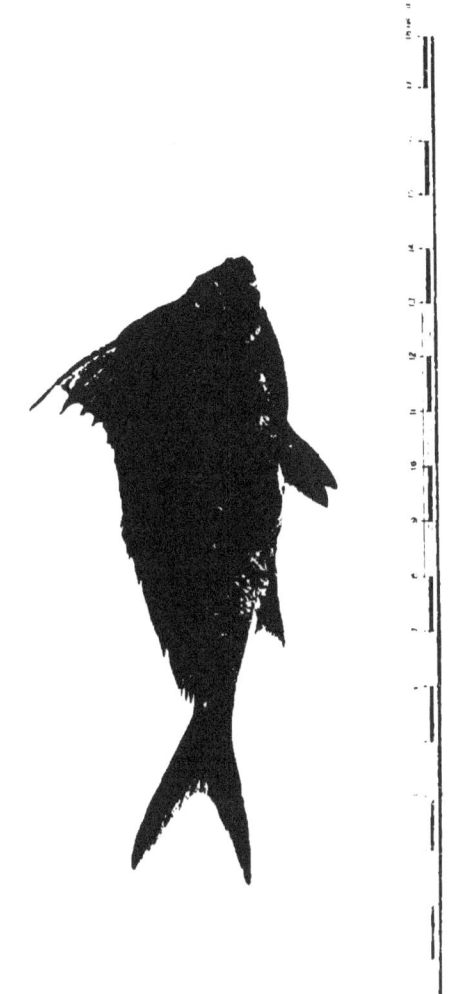

*Chiloductylus vittatus.*—GARRETT.
BANDED MORWONG.

rounded, distinctly denticulated; pectorals pointed, with a large skinny flap at the axil; soft dorsal with a scaly sheath at base; thirty to thirty-two rays, anal one spinous, eight soft rays; length reaching 3 or 4 feet.

The *Teraglin* is also a large and valuable fish, caught in the same way and at the same time as the jew-fish, but seldom reaching such a great size. The air-bladder of some of the same genus of fishes is of great value for isinglass, and forms a valuable article of export on the Indian and Malacca coasts, the merely dried bladder being worth equal to 3s. per lb. In our species—the teraglin—the air-bladder is of great size and excellent quality, and treated in the same way, that is without any preparation or outlay except that of drying in the sun, would probably add from threepence to sixpence to the value of each fish caught.—R.R.C.

The name Teraglin is stated by Mr. Hill to be an aboriginal one. The same author refers to a fish called the "Silver Jew-fish," which he considered from the form of the scales to be a different fish. Probably there was some mistake in the application of the native name, which it would seem Mr. Hill applied to the young of the true jew-fish, and his "Silver Jew" was the true Teraglin in its young stage. He adds that the "Silver Jew has all the appearance of one which would recommend itself; and certainly no one has as yet been deceived who ventured on the mere hypothesis of judging by appearance.* It is splendid when fresh caught, and also keeps pretty well if immediately cleaned after capture. They do not attain to a large size—3lbs. to 5lbs. weight. Silvery bright, and rapid swimmers, they love a good depth of water, and bite freely, now and then, at the ordinary bait. The tackle should be as fine as possible consistent with the strength of the fish, and the hooks well chosen. These fishes have much of the habit of their larger popular namesake, and frequently come in with the net. I doubt, however, whether the silver jew belongs either to the family or the genus of what is called or known as the jew-fish.

"It will be interesting to have these three fishes described, and from their particularly identical habits, one would almost imagine that they belong to the same family. The aboriginals, in whose sagacity I place great faith in such matters, have coupled two of these fishes only—mittila and teraglin. The silver jew, in all probability, has received its popular name from its similarity of habit."

## The Sword-fish.

Of the family *Xiphiidæ* or sword-fish, which have the upper jaw produced into a long sword-like weapon. We have one species which is occasionally seen at Port Jackson (*Histiophorus gladius*), Brouss. On July 4, 1880, a specimen was stranded at Wollongong nearly 14 feet long. The sword-fish is very dangerous to the schnapper fishermen. Mr.

---

* Mr. Oliver forwards me the following note on this subject :—"I do not think that the 'Teraglin' can be the same species as that popularly known as the 'Silver Jew.' Certainly the small silvery fish called 'Jews' caught in the Hunter River, Lake Macquarie, and the rivers and inlets of our coasts are not Teraglins. They may be, and probably are, the young of *Sciœna antarctica*, but in my opinion they are not *Otolithus*. The teraglin is a delicious fish, and is caught generally in offings or near the Sydney Heads. The little jew is caught up rivers, never in offings, and is wretched eating. I admit, however, that I have not compared them according to the scientific description."

Oliver states that on two occasions boats lying on the outer grounds were impaled by sword-fishes and the crews only saved their lives with the utmost difficulty.

## Fam. TRICHIURIDÆ.

Body elongate; compressed or ribbon-like; cleft of mouth wide; with several strong teeth in the jaws and on the palate. The spinous and soft portions of the dorsal and anal of nearly equal extent, long, many-rayed, sometimes terminating in finlets, caudal fin absent or forked.

These are marine fishes of very voracious habits, inhabiting mostly tropical or subtropical seas. It is either a surface fish or goes to moderate depths like *Berycidæ*.

## The Barracouta.

The "Barracouta" (*Thyrsites atun*) is the only valuable fish of this small family in Australian waters, but it is extremely rare in New South Wales. It is got in abundance in Bass's Straits and the Tasmanian coasts, and is sent in considerable quantity smoked and salted to the Sydney fishmongers.

This genus (*Thyrsites*) has the body rather elongate, without scales. The first dorsal continuous, with the spines of moderate strength, and extending to the second dorsal. From two to six finlets behind the dorsal and anal. Several long narrow sharp teeth in the jaws; teeth on the palatine bones. The species reach a length of 4 or 5 feet, and are valuable food fishes. The Barracouta, Barracuda, or Barracoota (and various other ways of spelling) is found from the Cape of Good Hope to New Zealand. In the latter place it is also known as "Snoek," and is exported from the Colony into Mauritius and Batavia as a regular article of commerce. *T. prometheus*, another species, extends from Madeira to Polynesia, and Dr. Günther thinks that our *T. solandri*, Cuv. and Val., is the same. The Tasmanian King-fish is *T. micropus*, Castelnau, one of the most delicious table fishes. It is much broader in proportion to its length than the Barracouta, and is caught at depths of 10 or 20 fathoms, off the south coast of Tasmania, in rocky places. The writer has had many a day's fishing off South Cape in Tasmania, and very frequently ere the fish was drawn to the surface it was bitten in half or sometimes the head only left on the hook by the numerous sharks which abounded in the locality. The Schnapper on our outer grounds is often brought in as a fragment, owing to the Blue Pointer Shark and other robbers.

The Barracoutas are well described as voracious fishes. They are easily caught from the stern of sailing-ships in Bass's Straits with no bait or only a piece of red rag. In 1857 the writer was on board a pilot schooner taking emigrants to Guichen Bay, S.A. Fair weather but foul winds kept us out to sea many days beyond our anticipated time and food began to get very scarce. Fortunately we came across an immense shoal of Barracouta, and for three days caught as many as we wished with nothing but a hook and line trailing fast astern. The quicker the vessel was going the better they used to seize the hook.

PLATE XIV.

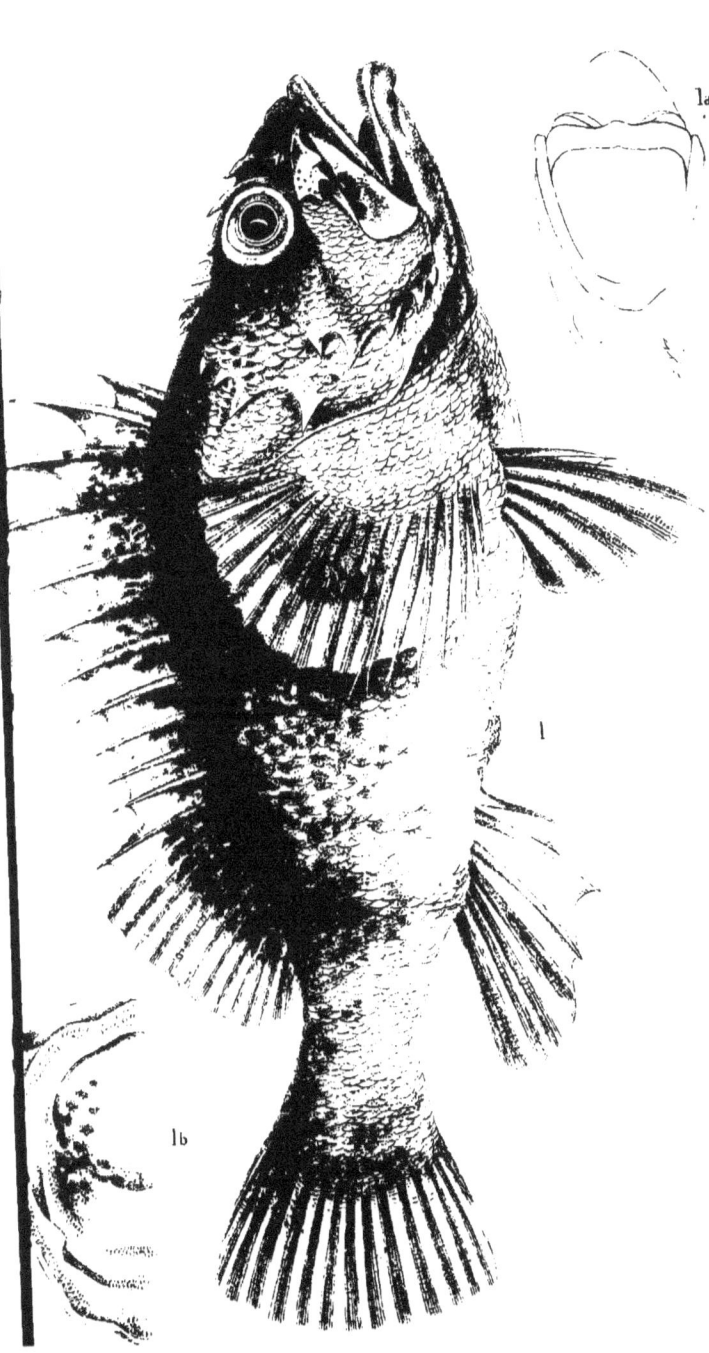

*Sebastes percoides.*—RICHARDSON.
RED GURNET PERCH.
1—Head, from front.  1ᵇ—Teeth.

In 1875 the writer went out with fishermen into D'Entrecasteaux's Channel in a small whaleboat with a fish-well in the centre. In the middle of the channel a fine shoal of barracouta was met. They seemed to be following a mass of fish spawn or young fry, of which we found quantities in their stomachs. We caught in a few hours as much as the boat would hold, probably some seventy or eighty fish from 3 to 4 feet long. Our mode of procedure was this :—Each had a small stout stick about 5 or 6 feet long and an inch thick ; to this was fastened a yard of log-line and at the end a square piece of cedar an inch thick and two or three long. The hook was fastened firmly on this so as to leave the barb projecting. A little piece of green hide was on the hook. All that we had to do was to splash and beat the water with the stick, pulling the line backwards and forwards, and after two or three turns if the movement was brisk and the splash considerable it was seized by a fish. The exertion of pulling it out of the water was great, as none weighed less than 6 lbs. and they were oftener 10 lbs. in weight. They had to be lifted clean up into the air and swung into the boat. We were always going fast through the water, and it may be supposed that the movement and the splash made a good imitation of the efforts of smaller fish when trying to escape from the remorseless *Thyrsites atun*. One can scarcely imagine more interesting sport or harder work than a day's fishing in this style for barracouta.

## DIV. XIII.—COTTO-SCOMBRIFORM ACANTHOP-TERYGIANS.
### Fam. CARANGIDÆ.

The first family in this division which need occupy our attention is the *Carangidæ* or Horse Mackerels, which are carnivorous fishes of tropical and temperate seas.

Body more or less compressed, oblong, or elevated, covered with small scales or naked, eye lateral. Teeth, none or conical. Spinous dorsal less developed than the soft or anal, either continuous with or separated from the soft portion or rudimentary ; ventrals thoracic or rudimentary or absent. No prominent papillæ near the vent. Gill opening wide. Ten abdominal and fourteen caudal vertebræ. Fishes of tropical and temperate seas.

Many of the fishes of this large family are in appearance and habit very like the more typical of the *Scombridæ*. There are very many species in Australian waters, their numbers increasing rapidly towards the warmer seas of the north. Those best known to the fishermen of Port Jackson are—the "yellow-tail" (*Trachurus declivis*), the "white trevally" (*Caranx georgianus*), the "king-fish" (*Seriola lalandii*), the "Samson-fish" (*Seriola hippos*), and "the tailor" (*Temnodon saltator*). The first of these, the "yellow-tail," is almost if not quite identical with the "horse mackerel" of Europe (*Trachurus trachurus*). In the young state it is abundant at all times in Port Jackson, and is in great demand for bait. The adult fish is seldom seen in the harbour, but is said to pass along the coast in large shoals at or about midsummer. It is most probable that this fish spawns in the inlets and harbours of the coast, from the fact that the young fish of 5 to 6 inches in length are always to be found in such localities. The very young fry have a most extraordinary and ingenious way of providing for their safety and nutrition at the same time ; they take up their quarters inside the umbrella of the large *medusæ*, where they are safe from their enemies, and are, without any exertion on their part, supplied with the minute organisms which constitute their food by the constant current kept up by the action of the curtain-looking *cilia* of the animal.

H

The "white trevally" is very abundant at times in the harbours and inlets of the coast, but generally in a young state. The adult fish is large, and appears in summer in very large shoals. The place of spawning is unknown. It is not much esteemed as a food fish.

The "king-fish" is about the most voracious and destructive of all the predaceous fishes of these seas. It grows to a large size, congregates in enormous shoals, and habitually pursues and destroys the shoals of other fish at all smaller than itself. It is not considered in its fresh state a very good fish, but when corned it is esteemed by some a great delicacy.

Of the "Samson-fish" very little has been observed. It is a large and handsome fish, affords good sport to the amateur fisherman, and is not much valued as food.

The "tailor" is well-known in Port Jackson. The young fish are constantly making their appearance in shoals in the summer season, and are taken in the seines in great numbers; they are much in demand for bait, but are not a favourite catch for the fishermen, as they are most destructive to the nets. The adult fish are large, and are known in the Melbourne market by the name of "skip-jack." They school in midsummer, move in enormous shoals, and are said to be most destructive to the young and spawn of other fishes. As an article of food they are not in much request, but when fresh there are few more delicate and well-tasted fish.—R.R.C.

A short description of some of these species is here given. *Caranx* and *Trachurus* are now included in one genus by Günther.

The body is compressed or nearly cylindrical, cleft of mouth moderate. First dorsal continuous, with about eight feeble spines or rudimentary. Scales small, curved in front, straight behind, entirely or posteriorly covered with plate-like scales, several of which are keeled, the keel ending in a spine.

## The Yellow-tail.

(Plate XVIII and Plate XXII.)

The "yellow-tail" of Sydney is *Trachurus declivis*, closely allied to the common British horse mackerel,[*] distinguished by having its lateral line armed with large vertical plates for its whole length, and a yellow tail; it is almost cosmopolitan in the temperate and arctic seas. It is also known by the name of scad. Mr. Hill says :—" It is not much good, except at sea along our coast, when occasionally it is better than salt beef, and is usually eaten when very fresh; it is dry in its character, and requires much garnishing to make it palatable. Like all deep-sea fishes, such as the bonito and albicore, it prefers a live bait, and is readily caught by an artificial one over the ship's counter. The pace these fishes swim at is astonishing. Coming coastwise recently in a steamer, and when off Port Macquarie, we caught several by towing a line over the stern, the hook covered with a piece of white rag in the shape of a small squid; then we were going about ten knots, and the mackerel appeared to be playing near the stern in the wake—every now and then they would start off as if the ship was at anchor. We were going too fast for the generality of fishes, that is to catch them in the way described. There are many of the deep-sea fishes caught in this way. The usual method is, when the vessel is going through the water at three to five knots to put out a line with an artificial bait, and to have a small bridle of twine

[*] Professor McCoy states that the Victorian *Trachurus* is identical with the Horse Mackerel of Europe. The figure given by him (Prod. Zool. Vict. pl. XVIII) is by his kind permission reproduced here (Plate XXII) from which it will be seen that it differs from the *Trachurus* of New South Wales.

*Beryx affinis.*—GŪNTH.
THE NANNYGAI.

fast as well as the line on board; the bridle being the first to have a strain is easily snapped when any additional weight is put on; consequently that being an indicator, soon tells when anything is fast. By this means a line can be set, and which requires little attention except when a fish is on, and does not bore the parties by looking after it." Young yellow-tails are caught in immense quantities in N. S. Wales as bait for schnapper. It is a fairly good bait, but not so good as mackerel.

## The White Trevally, No. 2.

*Caranx georgianus*, the "white trevally," is distinguished as a deeper fish in proportion to length, and the plates on the lateral line are little developed. There are several other species of *Caranx* in Port Jackson. In Victoria it is called silver bream. Count Castelnau says it is very beautiful when freshly taken from the water, the upper part being a light celestial blue or beautiful purple, the lower parts of a silvery white with bright iridescent tinges. Behind the operculum there is a black spot (also in the yellow-tail), and along the body extends a fine golden stripe. The dorsal fin yellow, bordered with black. There is another fish called by this name which has already been described amongst the *Teuthidæ*, but this is the White Trevally as generally known by N. S. Wales fishermen.

## The King-fish.

(Plate XIX.)

The king-fish of Port Jackson must not be confounded with the king-fish of Victoria (*Sciæna antarctica*) or the king-fish of Tasmania (*Thyrsites micropus*). This shows how confusing and misleading these local names are. Our king-fish belongs to a genus called "yellow-tails" in Europe. This is *Seriola lalandii*—Cuvier and Valenc. The generic characters are the same as *Caranx*, but the lateral line is not armed, and the body less compressed. In this species the colour is uniform, and the scales small, snout elongate, height little more than a quarter of the length, abdomen broad, not compressed, ventral fins moderate.

"The Australian king-fish," says Mr. Hill, "gives an idea of power and speed, when its beautiful symmetry and powerful tail are closely examined as it lies in the boat or ashore; but this idea is considerably enlarged when the amateur fisherman hooks one on a good fine line, with a determination to "hang on." That hasty resolution is quickly dispelled, and his idea more than realized as he finds the line tightening and whirring through his fingers, and any attempt to stop the fish at this moment would endanger these digits being either removed or cut clean to the bone, or lose that portion of the line already overboard. There are very few well practised fishermen who are desirous to catch more than half-a-dozen large king-fishes 50 or 60 lbs. weight in succession; they do a fair day's work when that is accomplished.

"The king-fishes appear in this harbour at regular and irregular intervals, but at all times waging a predatory warfare on other fishes, and often make bold to harass and shepherd them in shallow waters, making a dash and a splash among them at intervals, and which calls the attention of fishermen in search, when the net is speedily shot around them, and the depredators captured, together with their intended victims.

"I have known them in large numbers about the line boats, both between and outside the Heads, and along the coast, to rush up in mobs headlong to the surface at every occasion when a fish of another kind was hauled in; nothing would induce them to take a bait, and I have witnessed and participated in the sport, and have killed several with the spear when in their headlong and perpendicular career they have come up to the surface in the way described, but these were only comparatively small ones, and of moderate size.

"If you are fishing in the harbour, and the king-fishes are about, procure, if possible, a live yellow-tail or a mackerel, pass your hook through one of these alive, above the tail, so that it will not be disabled too much, put out your line that it may be able to swim away, and when all other bait have failed, the king-fish is almost certain to seize this. When the king-fishes are in good condition, and properly fresh, the belly part, cured and smoked, is far superior to any of the imported fish cured in that way: and that portion cured alone and used immediately, before it gets too salt, boiled and served up with egg sauce, is a choice *morceau*. Sometimes these fishes appear out of season, are lean and lanky; and then they are dry and comparatively ill-flavoured.

## The Samson-fish.

This fish has very minute scales, and is of a silvery colour, with a greenish back. It is much higher in proportion to its length than the King-fish, with a large head and a high short snout. Tail deeply forked, pectoral fins broad and short, ventrals black. This fish has five rather broad black cross bands and one above the eye, but they disappear in old specimens.

"The Samson-fish (*Seriola hippos*, Günth.) is occasionally caught either in the harbour or off the headlands outside; its habits are similar to those of the king-fishes, except that it moves about in deep water, and is more fond of the northern latitude.

"The great strength of these fishes is remarkable, and which probably is the cause that gave it the name of Samson-fish, as sailors or shipwrights give the name of a strong post resting on the keelson of a ship, and supporting the upper beam, and bearing all the weight of the deck cargo near the hold, *Samson post*. I saw one of these fishes caught in Botany Bay a few weeks ago, and it gave much trouble for its size, although I have seen them in good condition run away with a schnapper line and break them with ease."

## The Tailor.

### (Plate XX.)

*Temnodon* is the genus to which our "tailor" belongs. In this the body is oblong, compressed, covered with cycloid scales of moderate size; mouth wide, with single series of strong teeth, smaller ones on the vomer and palatine bones. First dorsal with eight feeble spines

PLATE XVI.

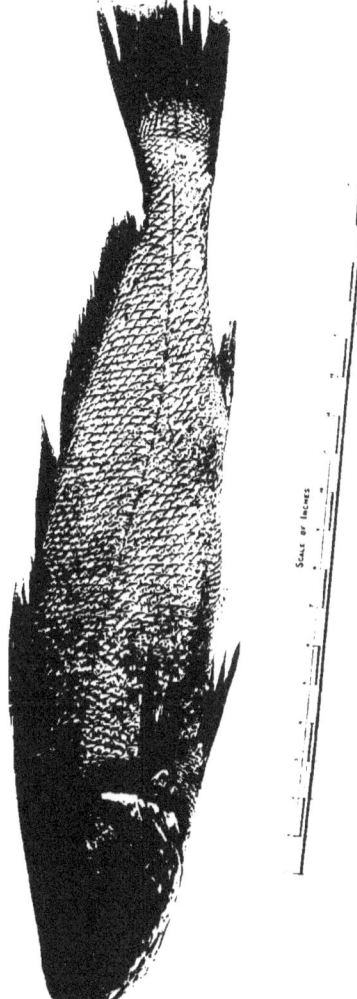

*Sciæna antarctica.*—CASTELNAU.
JEW-FISH.

connected by a membrane, no finlets. The second dorsal and anal covered with very small scales. Our species, *Temnodon saltator*, Cuv. and Val. (Plate XX), has a bluish, lead, or olive colour above, and silvery beneath. The maxillary bone reaches to a line drawn from the posterior margin of the eye, preoperculum with a notch above the angle and denticulations on the lower part. Dorsal spines very feeble. Of this species Günther remarks (G. S. F., p. 447) that it is found over all tropical and subtropical coasts, but is also met in the open sea. On the coasts of the United States it is known as the Blue-fish, being highly esteemed as food and furnishing excellent sport. It is one of the most rapacious of fishes, killing many more than it devours. It grows to a length of 5 feet, but the majority of those brought to market are not half that length (Günther). It is called skip-jack in Melbourne, which is also a name by which it is known in America and Britain.

Mr. Oliver informs me that this fish is very destructive to the fishermen's nets. A school of "tailors" enclosed in a seine generally involves wholesale destruction to the net. After such a haul a considerable expenditure of time and twine are necessary to repair the rents made by the sharp teeth of these very active and determined fish. Instances have been known where the entire bunt of a net has been torn to shreds by a few dozen sea tailors, whose vocation is the reverse of their land namesakes. The name "sea tailors" expresses the approach to the skip-jack size. Though Dr. Günther is the authority for tailors 5 feet long, 3 feet is an unusual size for those found in the South Seas.

The Bat-fish *(Psettus argenteus*, L., Plate XLV) belongs to this family, and is occasionally caught at Port Jackson.

## Fam. CYTTIDÆ.

Body elevated, compressed, covered with small scales, bucklers, or naked, teeth conical, small dorsal fin of two distinct portions. Ventrals thoracic. Gill opening wide. Marine fishes of temperate latitudes, the only genus of which is *Zeus*, or John Dory. Our species is *Zeus australis*, Plate XV, which Dr. Günther regards as identical with the *Z. faber* of Europe. The genus is distinguished by a series of bony plates along the base of the dorsal and anal fins, and another series on the abdomen. These fishes are regarded as excellent for the table. "The name seems to be a corruption of the Gascon 'Jau' cock, Dory being derived from the French, so that the entire name means gilt-cock." —G. S. F. The name of Gallo, which it has in other places, confirms this.

## The John Dory.

(Plate XXI.)

Among the other groups there is one fish of great value—the "John Dory" (*Zeus australis*). It is in every respect apparently the same as the John Dory of Europe (*Zeus faber*), and by some is regarded as the same fish. It is a ground fish, living on sandy banks and flats in moderately deep water. It enters Port Jackson in the summer months and is then full of roe. It is generally looked upon as a rare fish, but its rarity is probably owing to its being difficult to capture, the seine, even where its haunts can be got at, not taking sufficient hold of the ground to enclose it. The trawl would probably be found a more certain mode of capture. The excellence of this fish is universally admitted.—R.R.C.

They are caught very often with a line in April, in about three fathoms water in the harbour. The bait used is a small silvery piece of the side of a fish.

I have caught the dory at several places with the line and hook, and the moment it was dropped into the boat the mouth was pushed out to an enormous degree, as if indignant at the treatment it received. Singular to observe, I caught with the net a dory in Botany Bay at the time we were trawling for turbot. The dory has been long known, and when the currency of the Colony was in Mexican coin it was called a "dollar fish," and was esteemed a fine fish either as fried or boiled; but it is a rare fish, and may be considered a good one, though not from that cause.

I don't know how to describe the dory of the Colony; it is not a migratory fish, as we hear of it only now and then. It must be here, as Mr. Couch calls it elsewhere, a wanderer, following the fry of other fishes, on which, together with shrimps, it lives. The John Dory is known by name as well as any fish in the Colony from the scriptural allusion, and is easily identified from its peculiar mark on either side; that of this Colony bears all the characteristics, and those which I have handled weigh about 4 to 5 lbs.—E. HILL.

The Dory is of too greedy a temperament to like the short commons imposed upon those overgrown communities called shoals; he lives therefore very much to himself, frequenting such rocky sites as afford a safe retreat and an abundant supply of small fish.—Badham's Ancient and Modern Fish Tattle.

The families STROMATEIDÆ, CORYPHÆNIDÆ (Dolphins), and NOMEIDÆ need not delay us, as they are of no economic value to Australia.

## VII.—Fam. SCOMBRIDÆ.

Body oblong, scarcely compressed, naked, or with small scales, eye lateral, dentition well developed. No bony stay for the operculum. Two dorsal fins with (generally) finlets besides. Ventrals thoracic, with one spine and five rays. More than ten abdominal and more than fourteen caudal vertebræ.

This is the Mackerel family composed of ocean fishes of prey found in all temperate and tropical latitudes. They form, says Günther, one of four families of fishes most useful to man, the others being the Herring, the Cod, and the Salmon families. Their muscles receive a greater number of blood-vessels and nerves than other fishes, and are of a colour more like those of birds and mammals, and the temperature of the blood is warmer than other fishes by several degrees. Seven species are known of the genus *Scomber*, of which each coast seems to have its peculiar variety. It is distinguished by the feeble spines of the dorsal and the rows of little finlets behind the dorsal and anal, scales very small, and equally covering the whole body. Teeth small. Two short ridges on each side of the caudal fin.

## The Mackerel.*

With reference to our species, *Scomber antarcticus*, of Castelnau, Prof. M'Coy says that he cannot on comparison find the slightest difference between the Hobson's Bay and Mediterranean specimens. This would make our species *S. pneumatophorus*, De la Roche. It occurs rarely in Hobson's Bay, and generally in the month of June, and then in considerable numbers.

* An excellent figure of this fish, which is extremely like the European Mackerel, will be found in Prof. M'Coy's Prodrom. Zool. Vict., plate 28.

PLATE XVII.

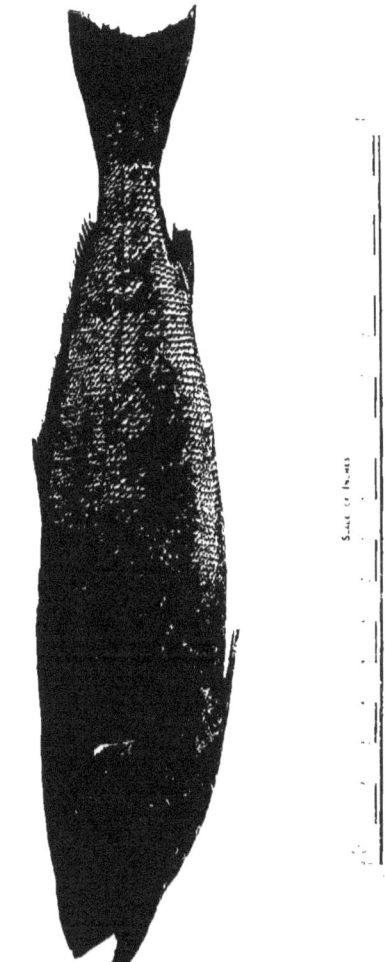

*Otolithus ateloitus.*—GÜNTH.

THE TERAGLIN.

The genus Scomber is represented in Australia by two species, *Scomber austral-asicus*, Cuv. and Val., and *Scomber antarcticus*, Castelnau. The last-named is the one best known in these seas as the "mackerel." Like all, or almost all of the "Scombrina," it is a gregarious and exceedingly predaceous fish, rejoicing in the open sea and generally near the surface, and apparently constantly in pursuit of shoals of other fishes. The instinct which in all fishes seems to compel them to mass together and approach the shore at the season of spawning is not wanting in the mackerel, and it is probable that the occasional visits of more than usually large shoals of these fishes is due to this annual movement; at the same time their frequent appearance in this harbour at unanticipated times may be owing to the appearance of shoals of other fish which they may have followed up. Be that as it may, it is certain that about midsummer, annually, shoals of enormous magnitude pass near the coast, going apparently in a northerly direction, that the sea has, sometimes for miles, the appearance of being almost a solid mass of them, and that they also have their enemies, and are followed and devoured by shoals of larger fish of their own family if not species, as well as by hosts of other predaceous fishes, birds, and mammals. The mackerel, which appear so frequently in Port Jackson, and afford such excellent holiday fishing to the citizens of Sydney, are generally very young, and it is said to be rare indeed to get an adult or full-roed specimen. It is a very good fish when eaten fresh, but like all the scombroid fishes it dies immediately after leaving the water, and decomposes very rapidly, and when eaten in that state it has been known to produce symptoms of fish-poisoning ; a most unjustifiable prejudice has been excited against the fish in consequence.—R.R.C.

The remarks of Mr. Hill on this family are so interesting and important that they will bear quoting at length :—" These are the well-known fishes (mackerel) of the Colony, of which everybody knows, and every fisherman has enjoyed the sport catching them, either late or early, in deep water or on the surface, biting freely, or very carefully and slow. Sometimes large, at others only half-grown, their name is legion ; and they visit this harbour three or four times a year, and remain longer or shorter as temptation offers in the shape of food, for they are very voracious, and live chiefly on young prawns or shrimps, stopping a week, or even a month, at each favourite bay in succession, so long as the food lasts. There you will see the boats congregated at dim dawn, catching mackerel with a hook and line. At other times these boats will remain the whole day, and during a cessation the owners will enjoy themselves as they think best. Sometimes at night, when travelling in shoals they will bite ; and if one can manage to break the school by catching a few he is sure to get many. The mackerel perform important services here —first of importance is providing bait at which every fish will bite, and at this alone when others fail. Besides they are readily caught when about, and are fit for immediate use. They also provide abundance of food ; but care should be taken as to when they are caught, and under what influence they have been placed since they were caught. We have no means in this Colony of catching them with a net in deep water, therefore it is not attempted, neither has this mackerel been accused of murder like it is alleged of a Norwegian, and is now a matter of history.

" A Norwegian author relates the story of a sailor belonging to a ship lying in one of the harbours on the coast of Norway, who went into the water to wash himself, when he was suddenly missed by his companions. In the course of a few minutes he was seen on the surface with vast numbers of mackerel fastened on him. The crew went in a boat to his assistance, and though when they got him up they succeeded with some difficulty in removing the fishes from him, they found it was too late, for the poor fellow shortly afterwards expired.

"I suppose this was from the overpowering numbers. There does not appear to be so many here on this coast, or in its harbours and bays, but they are numerous and migrate the same as they do on the coast of England; here, however, they are better disseminated, and do not move in column when they depart. At some seasons they are more plentiful than at others, like other fishes, and this well applies to Australia.

"In the whole of this family of fishes care should be taken when they are eaten. I recollect many times catching mackerel, and those for use I had their heads cut off immediately, and the body inverted in a bucket of water to bleed well, and no ill effects ever arose from eating them. At another time we were out for the night towards Manly Cove, and an immense quantity of mackerel were caught. The moon partly shone on them, and which produced a chemical change in their composition. Not only was the party in charge of the vessel made sick by eating them, but to whomsoever he presented them ashore they were made violently sick after partaking of them, although the man was cautioned about their then condition through the moon.

"The king-fish has also been the means, through this agency, of causing violent pains in the head, the face purple, and nausea of the stomach, and many instances of sickness and pains can be recorded, yet the people will not be sufficiently careful—they chance it.

"These fishes are excellent when fresh caught, but they will not stand long either the sun or the moon. They become flaccid and dangerous as food, and even as bait they are soft. When perfectly fresh they require a little more cooking than is usual with other fishes, and at the same time care should be taken that they are quite clean. It was thought at one time that the danger lay in the back-bone—it was consequently removed; others again thought it was that they were not divested at once of their entrails. All these experiments have been tried without effect. Of the fishes *Scombridæ* (which comprise those fishes just named, and also the bonito and the albicore, often caught at sea), decomposition, together with a chemical change, sets in at once, by the agency or through the influence of the moon, after which it is dangerous to eat them."

## The Bonito.

The bonito here referred to (*Thynnus pelamys*) is occasionally seen off our coasts, and also the king-fish of the West Indies *Elacate nigra*. We may also include the *Remora* or sucking-fish, in which the spinous dorsal fin is modified into an adhesive disk, occupying the upper side of the head and neck. It fastens on the shark and like the pilot-fish (*Naucrates ductor*) accompanies this predaceous fish in its rambles. They are both occasionally seen in Port Jackson.

## VIII.—Fam. TRACHINIDÆ.

Body elongate, low, naked, or scaly; teeth, small, conical; one or two dorsal fins, the spinous portion shorter and much less developed than the soft; anal, spinous and soft portions alike, no finlets; ventrals with one spine and five rays; ten or more abdominal and more than fourteen caudal vertebræ.

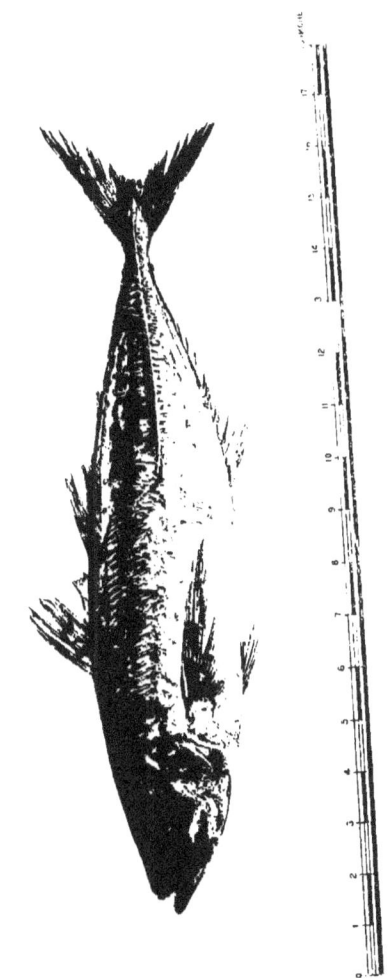

*Trachurus declivis.*—Cuv. & Val.

YELLOW-TAIL OR HORSE MACKEREL.

Carnivorous fishes of small size, in every shallow sea but more numerous in the Arctic than Antarctic circle. This family contains some remarkable genera which have the eyes on the top of the head and directed upwards (*Uranoscopus, Leptoscopus, Agnus, Anema*, and *Kathetostoma*). In the *Trachinina* division of the family the eyes are more or less lateral. This contains our sand whiting, *Sillago maculata*, Quoy and Gaim, Plate XVII. The body is covered with small ctenoid scales, the cleft of the mouth small and the upper jaw longer, eye large, lateral, two dorsals, first with nine to twelve spines, ventrals thoracic; villiform teeth on jaws and vomer none on palate; preoperculum serrated; head with wide muciferous channels.

## The Whiting.

(Plate XXIII, upper figure.)

The "whitings" are not like those of Europe.* There are in all four Australian species—the common sand whiting (*Sillago maculata*), abundant on the New South Wales coast ; the trumpeter whiting (*Sillago bassensis*), also abundant here, and the most common species in Brisbane ; *Sillago punctata*, the whiting of Melbourne, and rare on this coast ; and *Sillago ciliata*, Plate XXIV, occasionally seen here, and properly a fish of the tropical seas. The first of these whitings is far the most important as an article of food. It is perhaps in more general use even than the schnapper, constituting almost all through the year the most generally used breakfast fish we have. Of its excellent quality when in good condition and in the proper season there cannot be a doubt, but the great favour in which it is held induces, we fear, its extensive use at times when it ceases to be good, and may probably be unwholesome. It is in best condition when it first appears to come in from the sea, about the middle or latter end of summer. It is then a large clean fish, with the roe formed but not full-sized, and it continues in its excellent condition until the roe has attained full maturity or been shed. This occurs generally about March or April. The actual deposition of the spawn has never been observed, but there cannot be a doubt that the sandy and muddy beaches of bays and lakes are the favoured spots. There is a similar want of reliable evidence as to the time of the appearance of the young fry, but we believe that there are sufficient grounds for concluding that the spawn deposited at the end of summer does not germinate until the warmth of spring. The young fish, and those of all stages of growth short of the adult full-roed fish, seem to reside in the harbours, estuaries, and lakes in which they were born until their departure to the sea ; and it is while thus still, as it were, in their nursery that the most improvident havoc is played on them by the fishermen. The ages of these fishes at the time when they depart for the sea, and the precise period of their migration have not been determined by any accurate observation, but it is known that they invariably return in considerable shoals. From the evidence of Mr. C. Smith we are inclined to think that this whiting has two spawning seasons in the year, but if so, and there is little reason to doubt it, as it is certainly the case with the European herring and other well-known fishes—the spring spawning is much the least important. It is a ground fish, and, though it has been taken with the hook, is generally caught in the seine. The "trumpeter whiting" is not in such request as the other, nor is it found in such abundance. The time of arrival from the sea is winter, or a month or two later than the sand whiting. Its habits are much the same in other respects. The other species of *Sillago* need not be mentioned here, as they cannot be included among our useful fishes. There are other genera of *Trachinidæ* found in Australian waters, but none of them having any pretension to utility for food or any other purpose.—R.C.C.

In the sand whiting, which is shorter than *S. punctata*, the upper parts are of a light olive colour marbled with rather larger brown spots ;

* The whiting of Europe is a *Gadus* or Cod.

the lower parts are white; on each side of the body is a rather broad longitudinal band; the fins are transparent; and the rays spotted with orange; outer portions of caudal and dorsal dull, eye silvery.

Of the whiting, says Mr. Hill, "few will be found to dispute the fact that this fish ranks among the tip-top of the few choice and delicate fishes of this country; its very appearance indicates all that is good—beautifully clean, almost transparent, rigid and firm, and its flesh is as spotless as the driven snow; the fine delicacy of its flavour is nicely perceptible, and which renders it so valuable as an aliment to the invalid. The medical faculty, in recommending a fish diet to the patient for a change, generally prefer the whiting, for its delicacy, ease of digestion, and nutritive qualities.

"Fried or boiled whiting, served with suitable sauce as an adjunct to its flavour, is a charming dish for the more robust lovers of fish, and one which leaves not on the palate any coarse fishy redolency.

"The whiting commands the highest price generally, and is selected in preference by those who can afford the means of purchase. They are caught by net chiefly for the market, and the months of July, August, and September present the best harvest to the fishermen, at which time they go up the rivers in shoals of moderate size, and may be seen sporting on clear sandy beaches or in the deep channels of the estuaries, flashing their silvery sides to the light and making their presence known by the glistening which is thus occasioned. They are at this time in fine condition, and the cold season of the year offers an additional advantage to their own good keeping qualities.

"The whiting may be caught with hook and line off sandy beaches or sand-spits, and in open sandy bays on the coast in smooth water. The best bait is live earthworms, although they will also take fish bait, but not so readily; rareley however is the adult whiting caught by hook and line; the medium size is the rule, except on the open sea beaches, then some of the very large ones are taken by this means. The very young whiting, from 3 to 4 inches long, in the beginning of the year, and at early morning flood tide will readily take a worm bait. These, nicely cleaned, and fried crisp and brown, are not easily to be beaten, and would fairly vie with the famous whitebait of England.

"There is also another fish called rock whiting,* from the size and resemblance in shape to the whiting just described. They are frequently caught in the net, and are coloured with a greenish brown tinge and a few markings, but like many rock fishes are soft and ill-flavoured; if fried very fresh they are something better. A deal further south they are of firmer material, and hold a higher place in the estimation of fish connoisseurs."

*Kathetosma læve* belongs to this family, and is known as the stone lifter in Melbourne.

Passing over a number of families which are of no importance to Australian fisheries, we come to that named COTTIDÆ, which includes two genera of common Australian fishes. PLATYCEPHALUS or Flatheads, and *Trigla* or Gurnards. The name *Trigla* has been already spoken of as

* This is *Odax richardsonii*, one of the Wrasses.

PLATE XIX.

*Seriola lalandii.*—Cuv. & Val.

THE SYDNEY KING-FISH.

signifying its habit of spawning thrice a year; but Athenæus affirms that this species only breeds three times in all. The tradition is worth bearing in mind.

## Fam. COTTIDÆ.

Body oblong, subcylindrical, mouth-cleft lateral, feeble teeth in villiform bands. Head with spines and a bony stay connects the preopercular spine with the infraorbital ring. Two dorsal fins (rarely one), spinous less developed than the soft or the anal. Ventrals with five or less than five soft rays.

These fishes are small, remaining generally on the bottom in shallow water.

## The Flathead.

(Plate XXV.)

PLATYCEPHALUS is easily distinguished by its depressed head, and body which becomes cylindrical towards the tail. About forty species are known, of which some attain the length of 2 feet. They live on the bottom in shallow water, hidden in the sand, the colours of which their bodies resemble. They are scarce near islands or deep seas, but the number of species is considerable in the temperate portions of the Australian coast, where the bottom is sandy. "Their long and strong ventral fins are of great use to them in locomotion."—Günther.

The "flatheads" of the coast are *Platycephalus fuscus, levigatus, bassensis*, and *cirronasus*. Of these, the first, *P. fuscus*, is the flathead best known and most common in Port Jackson. All the species are of excellent quality, and may be ranked amongst the best of our fishes. Like the red rock cod the flathead is a ground fish, but is found on a sandy bottom only, and generally at only a moderate depth. It is taken both by the hook and net. But little information is attainable about the history of the flathead, but it seems to come into this harbour full of spawn in midsummer, and probably deposits its ova on the sandy banks in tolerably deep water.—R.R.C.

Some of the species of flathead will venture up the rivers into fresh water. Thus *P. fuscus* comes up the river Hunter as far as West Maitland, where it is caught abundantly by the anglers in summer.

Of *Platycephalus fuscus*, Cuv. & Val., Mr. Hill says:—"They are ground fishes and bite freely in the summer season, but retire into deeper water during the colder months of the winter, where they might also be caught with the line. Of a calm day it is usual to let the boat drift over the ground, which is generally sandy, when occasionally they bite freely. I have often known fifteen or twenty dozen hauled up, and as fast as the line could be put over. They are sluggish fishes, and do not give much sport; the very large ones of 3 or 4 feet in length, which are at times caught with the line, have pretty good strength, but are soon exhausted. The flesh is good, white, firm, and flaky, and it is preferred when boiled. Some consider the flathead a fine fish. Certainly its outward appearance when dead does not recommend it.

"The flathead is armed with a double thorn over each gill operculum, and which give a bad laceration, afterwards accompanied with pain. Care must be taken in removing them from the hook that these bones do not come in contact; the usual plan is to give the fish a tap on the head with a short club for the purpose, hard or gently, according to the size. These bones appear to be lifted at will, to show the anger of the fish, and are about from 1 to 1½ inch long.

"These fishes take a bait freely at night. I have seen the deck of a vessel strewn with flathead after a couple of hours' fishing. Three or four vessels had taken shelter in Botany Bay, on the south side, from a southerly gale which prevailed. The fishes also appeared to have moved over, from the same cause perhaps, as the place was literally alive with them, and many were captured with the hook and line on that occasion, but no very large ones were caught; they appeared to be all young fishes about 18 inches and some 2 feet long."

## The Flying Gurnard.

*Trigla*, or Flying Gurnards as they are called, are well known from their peculiar squarish heads and their long and broad pectoral fins like wings. One of their peculiarities is the possession of three finger-like pectoral appendages, which serve as organs of locomotion as well as touch. They make a grunting noise when taken out of the water, which is, says Günther, from the escape of gas from the air-bladder through the pneumatic duct. The colours of the pectoral fins are very beautiful, as the following description of our species, *Trigla kumu*, Lesson (plate XXVI, upper figure) will show. It is taken from the Prodromus of the Zoology of Victoria, by Professor M'Coy, F.R.S., who gives an excellent coloured figure:—"Dead yellowish on cheeks, purplish grey above on front and behind, back and sides a dull pale cinnamon brown, with an olive tinge and with large blotches, irregular in size, shape, and disposition, of a dull Indian-red or reddish brown. Dorsal fins very pale brownish and yellowish, anal and ventral with reddish rays on the outer side, on inner side pale yellowish olive with a pale narrow greyish blue margin, with two or three rows of small oval spots of the same colour within the border. A large ovate black blotch between the third and sixth rays from the lower edge, having about five oblong, opaque, white spots, tinged with bluish along its upper edge, and three running down the middle; caudal fin reddish, with browner rays darker towards the margin; throat and belly pure white, iris golden yellow." We have two other species in Port Jackson, besides species of the nearly allied genus *Lepidotrigla*, only distinguished by larger scales, which Dr. Günther now unites with *Trigla*. "Resplendent in their covering, brilliant in their ornament, rapid in swimming, swift in flight, living together without strife, defending without injuring themselves, one would think they should be included amongst those beings on which Nature has bestowed the most favours."—Lacepede.

[In plate XXVII a figure is given of *Trigla polyommata*, Richardson, or the flying gurnet, which is found on all the Australian coasts, from New South Wales to Western Australia.]

"The gurnards," says Mr. Hill, "also rejoice in the name of growlers, from a sort of grunt which they give when taken out of the water. Perhaps the present *Trigla* is exempt from such a stigma—at least it is in the waters of Port Jackson, so far as I know. Certainly it is one of the most beautiful frequenters of our harbour, and singular for the size

PLATE XX

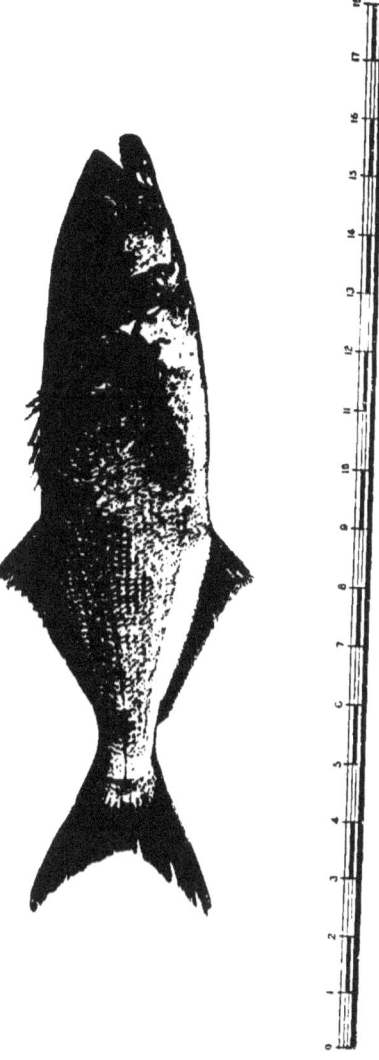

*Temnodon saltator.*—CUV. & VAL.
THE TAILOR OR SKIPJACK.

of its pectoral fins, which would lead one to imagine it capable of supporting its body in the air, like the noted flying-fish. It is not, however, known to exert this power, or even to possess it. Their heads are said to be so placed as to enable them to move among the stones and rocks, and remove from thence any crabs or shell-fish which may be concealed, their well-armed head preventing them from receiving injuries while exploring the rocky retreats for their prey. The flying gurnard is not a common fish here, and its flesh is not so dry as in gurnards of other parts of the world, and, served up with a nice sauce, it makes a very good dish."

The family *Cataphracti* can be passed over, merely stating that one species, *Dactylophorus orientalis*, Cuv. and Val. (which has the habits of the Flying-fish), is seen as far south as Port Jackson, though more commonly in the tropics. It is the Indo Pacific species.

The families of "Gobies Blennies," &c., have been already mentioned. Passing at once to the eleventh division of the Ancanthopterygians, or those generally resembling the mullets. They have two dorsal fins more or less remote from each other; the anterior either short, like the posterior, or composed of feeble spines. Ventrals abdominal, with one spine and five rays.

## SPHYRÆNIDÆ.

The first family is the Barracudas, or Sea Pike.* There are three fishes of this small family occasionally found in Port Jackson and on other inlets of the coast. They are easily known by the elongate muzzle and strong teeth, and in general outline are not unlike the British pike. This gives them the name which they bear of "pike" amongst the Sydney fishermen. They are only caught by the seine, and are most abundant in summer, when they seem to come to the coast in small shoals. They are also caught at other seasons, but the adult fish rarely in large quantities. The young are taken much more frequently in seines.

### The Pike.
(Plate XXVIII.)

"Pike," as generally understood here, is *Sphyræna obtusata*, Cuv. and Val. (Plate XXVII), a long narrow fish, with a long head and scaly opercula. It is greenish lead colour above, silvery beneath, fins deep yellow, outer rays black, a dark band from snout to tail below lateral line; there is also a black blotch under the root of the pectorals. We have another species in Port Jackson, *S. Novæ Hollandiæ*, Günth. Another kind of Pike is *Lanioperca mordax*, Günth. (Plate XXVIII).

### Mugilidæ or Grey Mullets.

Distinguished by moderately large cycloid scales, dentition very feeble or none, no lateral line.

---

* This name is no doubt the same as Barracouta, and is of Spanish origin. The application of it to *Thyrsites atun* in the Southern seas was founded on some fancied resemblance to the West Indian fish, which originally bore the name, though of course they are entirely different.

The grey mullets, says Dr. Günther, numerous in species and individuals, inhabit all the coasts of temperate and tropical climates. They frequent brackish waters, where they find their food, which consists of organic substances, mixed with sand and mud; to protect their stomachs and gill-openings they have the pharynx modified into a filtering apparatus. They take in a quantity of sand and mud, and having worked it for some time between the pharyngeal bones eject the useless portions. These bones are coated with a thick soft membrane, which rests upon a large fatty mass, giving it considerable elasticity. Another oval mass of fat occupies the roof of the pharynx between the two pharyngeal bones. Each branchial has on each side a series of closely-set gill-rakers, which are bent sideways and downwards, each set closely fitting into the series of the adjoining arch, thus constituting an admirable sieve. "Some seventy species of grey mullet are known, the majority attaining a weight of about 4 pounds. All are eaten, and some are esteemed, especially when taken out of fresh water. If attention were paid to their cultivation, great profits could be made by securing the fry and transferring it into suitable backwaters on the shore, in which they would rapidly grow to a marketable size."—G.S.F., p. 504.

Two genera only will occupy us, and they are easily distinguished. *Mugil*, without teeth; *Myxus*, with feeble teeth.

The best known species in New South Wales are the "sea mullet" (*Mugil grandis*, Castelnau), the "flat-tail mullet" (*Mugil Peronii*, Cuv. and Val.), the "river or hard-gut mullet" (*Mugil dobula*, Günther), and the "sand-mullet or talleygalann" (*Myxus elongatus*, Günther). Other species there are, such as *Mugil cephalotus*, *Petardi*, *compressus*, *argenteus*, and *acutus*, but they are rarely seen and little known, and not therefore classifiable as useful fishes. The first of this list, the sea mullet, is a large fish, attaining when full grown a length of 2 feet and a weight of 8 lbs. It is unsurpassed in richness and delicacy of flavour by any fish in the world, the salmon not excepted, and it offers itself for our use in countless numbers at the very season when it is in the best possible condition. The history of this fish is now pretty well known, though it will be seen by a perusal of the large amount of evidence printed in the Appendix that there are many very conflicting statements and opinions given.

To begin with the spawning season:—In the latter end of summer, that is at periods varying from the middle of March to the middle of May, the sea mullet is seen to enter all the harbours and inlets of the coast in successive shoals, some of the most astonishing vastness. It is then full of roe, and in splendid condition. When not interfered with by fishermen (for it is a fish easily turned from its course) or diverted by storms or floods, these shoals penetrate to all parts of these inlets, and run up the rivers even into fresh water in search of suitable places for the deposition of their spawn. When a suitable spot is reached, the deposition of the spawn commences, and the process is carried on in much the same way as that of the salmon and other fish of similar habits. Sometimes, however, from bad weather or the persecution of fishermen, the shoals are prevented from seeking suitable spawning grounds, and the fish being no longer able to retain the spawn, shed it loose upon the water, where it becomes entirely lost. When the ova are properly fertilized and left undisturbed, the young fish make their appearance on the approach of warm weather in spring, when they may be seen in large shoals close to the land and in shallow water. From that period until they become adult, which is probably at the age of two years, they seem to keep entirely to the rivers, lakes, and mud flats, where they thrive and grow with amazing rapidity. When in this half-grown condition they are very inferior in flavour to what they become afterwards, having an oily and muddy taste. As they are without teeth, they are incapable of eating either animal or vegetable substances in the ordinary sense of the term, but they are possessed, Dr. Gunther informs us, of a pharyngeal apparatus which sifts the organic from the inorganic

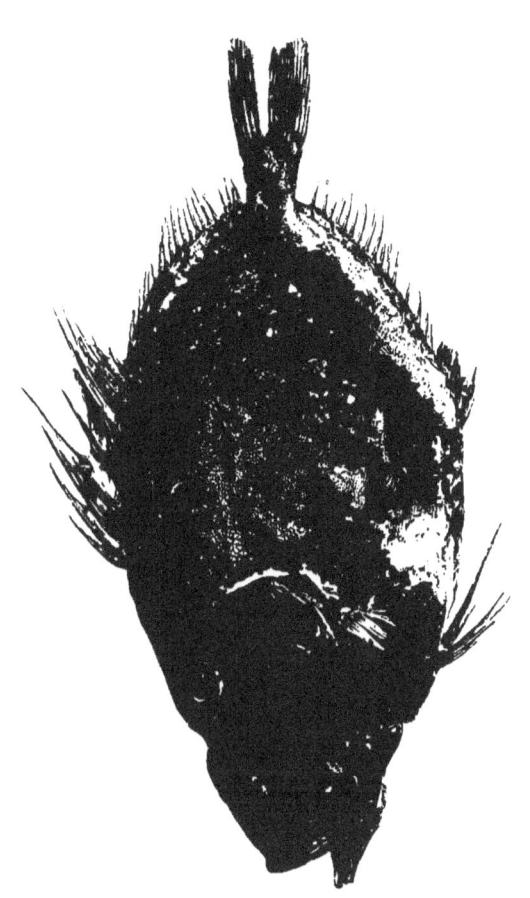

particles from the mud which they swallow and on which they live. When the period at length arrives for the mature fish to go to the sea preparatory to spawning, the instinct which actuates them seems to be irresistible. In one instance some years ago, when Tuggerah beach lake was for a time shut up at its sea mouth, the mullet pressed in such masses in the direction in which the outlet should have been that thousands of them were forced up on the land and perished. An occurrence of the same kind is mentioned as having happened at Lak Illawarra under similar circumstances. It is doubtful how long it is between tr rush of the fish to the sea and their re-entrance into the same or other rivers ; the belief is that the time is very short, that the movement is only from one opening of the coast to another, and always from south to north. There can be little doubt that the fish after spawning find their way back to their old haunts, but they have very seldom been seen so returning. The spent fish are for a time unfit for food, but they improve in condition very rapidly. The only instrument of capture used for the mullet is the seine net. The range of the species is from the Gippsland lakes on the south up to Brisbane on the north.

The flat-tail mullet is also a very good fish, but has neither the size nor the extreme excellence of the sea mullet. It appears also at the end of summer or beginning of winter and spawns in our bays and creeks, but the shoals are never of the same enormous size as are common with the other.

The other species (*M. dobula*) is, except at the schooling season, almost a freshwater fish, living as high up the streams as it can get, but it cannot, like the European salmon, pass up rapids or falls. It is a good fish, but inferior to both the others. The term "hard-gut mullet" is sometimes applied to this species, but more frequently the fishermen apply that name to immature specimens of the "sea mullet." It is sometimes taken by the hook.

The "sand mullet" (*Myxus elongatus*) seldom exceeds 7 or 8 inches in length, and, though no doubt excellent eating as are all the family, is looked upon as too small for the market.—R.R.C.

## The Sea Mullet.

(Plate XXX.)

A figure of *Mugil grandis*, Castelnau, is given as it is a fish which may become of more commercial importance than any other. The observations of Mr. Edw. Hill are here inserted, with a little alteration.

"There are several varieties in this Colony, but the one to which especial reference is made may be readily identified by its size and obesity, as also by the numbers which are brought into the market at a time, by boat-loads and cart-loads, and usually at this particular season, when they are so much needed. These fishes come from the south, and go as far north and east as Navigator Islands, and enter the bays and harbours of this coast from February to April.* In March, however, they visit Botany Bay, Sydney Harbour, and Broken Bay, sooner or later in this month, and as a rule may be looked for after every southerly gale or easterly bad weather. As these fishes are then full-roed, no doubt they are migrating for the purpose of spawning, and keep together in large shoals, at times covering acres, and when they do arrive great attention is bestowed on their movements by professional fishermen, both late and early, and if a chance is offered by their coming into water sufficiently shoal they are surrounded by a net, and sometimes double-banked by net after net, at which times thousands are captured. These fishes at

* This is not quite correct. The Australian species have only been seen between Gippsland and Brisbane. The one referred to as seen off the Pacific Islands is another species. The fish cannot live long without the mud from which it derives its food.

this juncture are the poor man's friend, being cheap, rich, palatable, and wholesome ; and were it not for the fact that they offer too abundant a feast during this short season they would be considered one of the most useful fishes of the Colony, and would provide us with large store for the future were advantage taken of the great catches which are so frequently made. Salted mullet and smoked mullet are not unknown, and they are excellent too ; and why should there not be kippered and pickled mullet also ? But the fact is they come and go so very soon that their presence is scarcely recognized till they are nearly gone. During that brief season what myriads could be captured ; five or six boat-loads have been caught at one haul with the seine ; tons upon tons have been surrounded with and the net staked up to keep them from leaping over or from fear that some of them would get on the cork line and keep it down. Most assuredly at such a time the rush would be so great, and a general follow-my-leader take place, like a flock of sheep, that it could rise no more till the last had passed.

"The great weight of this multitude of fishes in the net when it is called a large haul necessitates the cutting of a wing from the net wherewith to sweep ashore from the great bulk what might be required. At one time there was no difficulty in getting cured mullet ; experiments have also been tried by boiling them down, when it was ascertained that each mullet yielded nearly a pint of fine clear oil. These fishes are very fat, and independent of what is contained disseminated throughout the body there is one solid piece, a huge flare, a kind of magazine, wherein the fish draws nourishment during its migration, and which comes away separate and yields a large proportion of the oil. Boiled mullet, soused mullet, or choice fresh fillets fried, are very good, and in my estimation the 'sea or sand mullet' during their season are not excelled by any of the fishes of the Colony. The question might be asked, 'what becomes of those not sold ?' and may be as readily answered, thrown away to drift out with the tide, or to feed the sharks. At one time they were used as manure, but now that appears to be too much trouble. A few boat-loads stall the market, and there is no outlet for the surplus ; the fisherman's harvest, as it used to be called, has lost its name and prestige. During the migrations of these fishes they are followed by large sharks, as well as by other carnivora, and when they come into shallow water great rushing about is caused by these attendants. These mullet will not take a bait, and the means employed to capture them is by net. The aboriginals used to build weirs in the mouths of small creeks, when they found that the mullet had gone up, and used their spears to procure sufficient for a greasy banquet ; the feast was theirs, and there they remained for weeks till surfeited with the gorge. It has been argued, and perhaps with some degree of truth, that these identical fishes go high up the rivers and creeks at spawning season to deposit their ova in mere brackish water, and in due course, when such ova comes into existence, the young fry remains. Certain, however, in fresh water, high up the Hawkesbury, and in the eastern and south creek tributaries to the same river, adult fishes, with all the characteristics of the sea or sand mullet, have been caught by nets placed across the stream or the more still waterholes and reaches of the creeks. These fishes will not take a bait, and it was usual for the aboriginals, when a net had been so placed, to go into the water above that position and drive them towards its meshes.

PLATE XXII.

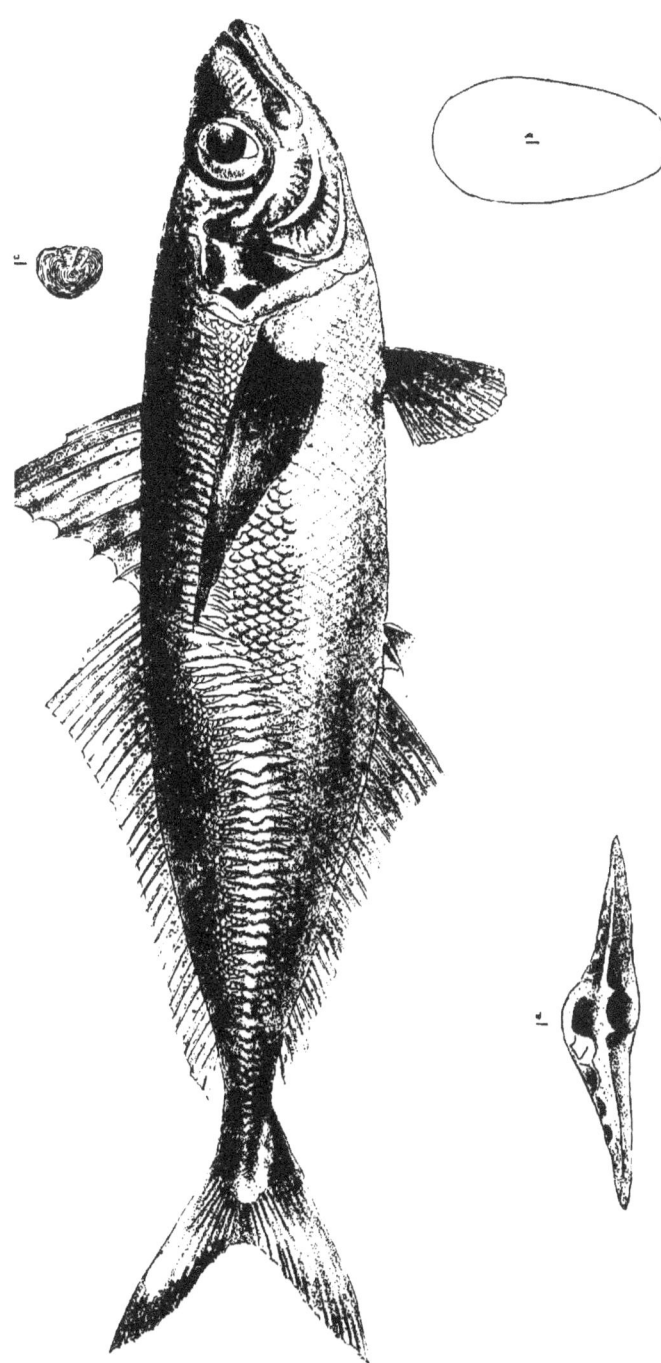

*Trachurus trachurus.*—Cuv. & Val.
HORSE MACKEREL OF VICTORIA.
1ª—Armour plate of lateral line. 1ᵇ—Section of fish. 1ᶜ—Scale.

"Mugils are very distinct from the sea or sand mullet, besides which they come down the rivers, and go in shoals or schools in their season, but do not attain more than one-third the size of the sea mullet, and are frequently brought to market with the ordinary net fish, and by which means they are usually caught. They are tolerably good and palatable fishes, but not so estimable as the large kind named, and it is curious and instructive to see practical fishermen following these shoals of fishes for the purpose of capture. Almost instinctively they will tell you whether they are the hard-gut or the flat-tail, or the tallagallan, directly they see any one of these leap from the water, a mode of procedure very common to this family of fishes when sporting uninterruptedly. The "*hard-gut*"* is the best marketable fish of these three, and may be distinguished by his short thick tail, flat deep sides, and shining scales. The *tallagallan* is a long-bodied fish, and found chiefly on the flats at flood-tide, or near shoal water with a deep stream or channel in one part. In the young state these fishes are much more delicate and palatable than the adult. When they are dry they form, however, a very good adjunct to the other fishes of the Colony. The flipper or flat-tail is found at the mouths of small fresh-water streams on the clear white sand, and readily take a bait of dough and bread-crumbs kneaded together, and at which times they afford good sport to those who delight in the pleasure of fishing with the rod and line, but they are not more than one-third or half-grown at this time."

The sea-mullet affords good sport to anglers in the Hunter and other eastern rivers. The bait is a small worm, but a far better kind is the fine silky green conferva which grows on the surface of stones or logs which have been long in the water. Mullet will take this with great avidity. The weed must be cut rather long and wound around the hook. It must not be confounded with the coarse green woolly conferva which covers the bottom like a blanket. Mullet will not touch this. The other they eat so greedily that not a particle can be found on the stones and logs of the rivers where these fish are abundant.

A good many divisions and families of fishes must be passed over to come to those of most importance to New South Wales.

## II.—Ord. ACANTHOPTERYGII PHARYNCOGNATHI.

Parts of the rays of the dorsal, anal, and ventral fins are non-articulated spines. Lower pharyngeals coalesced, air-bladder without pneumatic duct.

### POMACENTRIDÆ, or CORAL FISHES.
### LABRIDÆ, or WRASSES.

The Wrasses are a large family of littoral fishes, very abundant in the temperate or tropical zones. Many of them are readily recognized by their lips, which are sometimes internally folded, a peculiarity which has given the name of *Labridæ*, from *labrum*, lip. They feed chiefly on molluscs and crustaceans, their teeth being admirably adapted for crushing hard substances. Many species have a strong curved tooth at the posterior extremity of the intermaxillary, for the purpose of pressing

* The "Hard-gut" is merely an early stage of the Sea Mullet. A. Oliver.

a shell against the lateral and front teeth by which it is crushed. Others feed on corals or zoophytes, and some are herbivorous. Beautiful colours prevail in this family, and some species which are the most prized as food reach a weight of 50 lbs.

This family includes all the rock and reef fishes known as parrot-fishes. A large number of species are found in our seas, but many of them are only occasional visitors from the warmer regions of the north, where the *Labridæ* abound. Those that are most familiar to the Sydney public are the "blue groper" (*Cossyphus gouldii*), the "pig-fish" (*Cossyphus unimaculatus*), the "Maori" (*Coris lineolatus*), and the "rock whiting" (*Odax semifasciatus*). The first of these is a large fish, and though very little appreciated is exceedingly good, indeed the head makes the most delicious dish one can well conceive. It is occasionally taken in the seine when making a haul near rocks, but the usual way of capturing it is by a spear, a mode of fishing which the aborigines and their half-caste descendants are very expert at. Several other species of the *Labridæ* are said to be very good for food, but they are little known.—R.R.C.

*Cossyphus* here referred to has a compressed oblong body, with scales of moderate size, imbricated scales on the cheeks and opercles, basal portion of vertical fins scaly, lateral line uninterrupted, teeth in jaws in a single series, four canine teeth in front of each jaw, a posterior canine tooth.

## The Blue Groper.

(Plate XXXI.)

*Cossyphus gouldii*, Richardson, or the blue groper is a very large dark purple fish, attaining a length of from 3 to 4 feet. It has no posterior canine teeth, the scales on the cheek are not imbricate, dorsal fin scaly, and the caudal truncate.

Mr. Hill speaks of this fish as the "gruper," and says "it is popularly called in this country the blue or black groper, no doubt from the fact of these fishes groping in and out of the caverns and crevices of rocks in search of crustacere.

"The gruper, though plentiful, is not a common market fish, neither is it sought much after by professional fishermen, for various reasons, among which may be mentioned that it is out of their lay for their general fishing grounds, as it is essentially a dweller among the caves and rocks of the coast, and the rocky points within the harbour, where their occupations rarely extend, and also that the gruper may not be, from its coarse appearance, a favourite with the general public.

"It is a clumsy-looking fish, with huge scales which hang on with great tenacity, requiring a tomahawk to remove them, unless it is done immediately after capture—even then it is a tough job. These fishes grow to a large size, and attain a weight of 30 lbs. or more, either blue or bronze when in the water; when taken and dead they soon change to a very dark brown or black. These fishes carefully skirt the rocks at flood tide in seach of food; on the ebb they retire at once to deep water, and of a calm day may be seen at a moderate depth sporting leisurely near the vestibule of their own rocky halls.

"The gruper is a shy fish when he sees moving objects on the shore. The aboriginal when seeking this fish, armed with a couple of spears, prosecutes his search with cat-like caution, and when in view is as motionless as a statue, keeping, if necessary, that position for a considerable time till a chance offers, when he darts one of the spears with

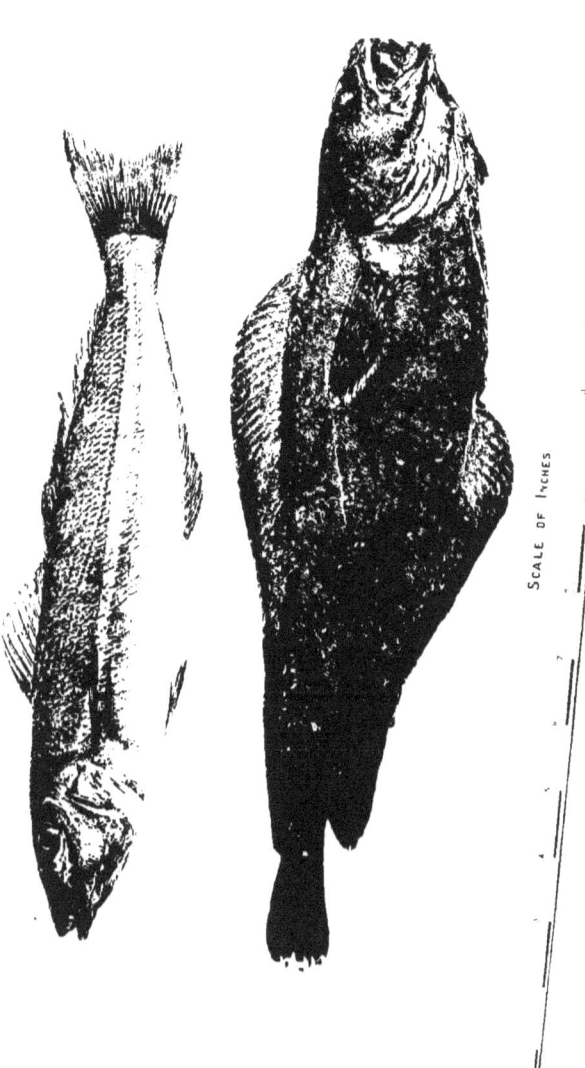

Sillago maculata.—Q. & G. (upper figure).
SAND WHITING.
Lotella marginata.—MACLEAY (lower figure).
BEARDY OR COD.

an unerring and powerful plunge into one of the fishes, and fixes it firmly to the rock or ground. Then the struggle commences; the second spear is brought to bear, and in general accomplishes the work of death.

"The gruper is often decoyed from its hiding place by oysters and shells broken, or crabs broken up and thrown out as berley; then the cunning and dexterity, added to which the patience of the black is admired. There is nothing to equal his agility when he boldly pursues from rock to rock the fugitive, perhaps with a broken or disabled spear, and at length dashing in with his already disabled shaft he will struggle hard and do his work effectually.

"With hook and line along the rocks of our sea-coast these fishes are caught, but the bait should be crabs. It is usual to wrench legs and shell off the back, and cast them out for berley. The body is then secured on the hook with a piece of thread or flax and thrown out; if a gruper be at home, that is the surest way to entice him.

"Boiled head or shoulder of a gruper is a perfect dish; and I have learned from those who are not only capable of judging but giving an opinion also, that it more resembles the princely turbot than does any other of the fishes of the Colony."

### The Pig-fish.

Another species called the pig-fish, *C. unimaculatus*, Günth. (Plate XXXII), is a much smaller fish, with the pre-operculum minutely serrated, and exceedingly strong anal spines. It is of a uniform red colour, with a deep black spot at the base of the sixth and eighth dorsal, and a small black spot on the fifth and ninth dorsal spines.

### The Rock Whiting.

*Odax*, "rock whitings," have a compressed oblong body, with small scales, head naked, lateral line not interrupted, dorsal spines nine. We have only one species in Port Jackson, already referred to.

### The Stranger.

*Odax* has a conical snout with the edge of each jaw sharp and cutting without distinct teeth. The Port Jackson species named is about 14 inches long, of a sky-blue colour, with a golden spot on each scale, or green with the belly white, and with transverse black spots on the back. Another species is called the "stranger" in Melbourne (*O. richardsonii*, Günth.) and another in Tasmania, *O. baleatus*, Cuv. and Val., goes by the name of "kelp-fish." A very small one in Port Jackson, of chocolate brown colour with reddish fins, is called *O. brunneus* (Macleay).

## III.—Ord. ANACANTHINI.

Vertical and ventral fins without spinous rays; ventrals jugular, thoracic or none. No pneumatic duct in air-bladder if present.

### GADIDÆ, or COD-FISHES.

The cod family, so largely and usefully represented in Europe and America, only exists in Australia in the form of two species of *Lotellacullarias* and *fuliginosa*, known to the fishermen by then ames of "beardie" and "ling." They are very rare, and generally of small size. Nothing is known of their edible qualities.—R.R.C.

## The Beardie, or Ling.

(Plate XXIII, fig. 2.)

LOTELLA MARGINATA, Macleay (Plate XVII, fig. 2) belongs to a genus which has a separate caudal fin, teeth in the upper jaw in a bad and an outer series of larger ones; chin, with a barbel. There are four species known in Australia. Our Port Jackson species is from 14 to 20 inches long, of a uniform brownish colour with the margins of all the fins white. Of the esculent qualities of the species nothing is known, but those in Victoria and Tasmania are valued by some. The flesh is always very soft and watery. There are excellent figures of the Victorian species in McCoy's Prodromus of the Zoology of Victoria, Part II, plates 19 and 20, caught with a line off the rocks; the fishermen state that the spawning time in Victoria is April.

Passing over the families OPHIDIIDÆ and MACRURIDÆ, the second division of the Anacanthini demands a more lengthened notice. It consists of one family—

## PLEURONECTIDÆ.

These fishes are called "Flat Fishes," including soles, flounders, plaice, turbot, &c. They have no air-bladder, and are so compressed that they move on one side of the body. The lower side, which on different genera is sometimes the left and sometimes the right, is white, the upper side is variously and sometimes brilliantly coloured, and both eyes are on the upper side, an arrangement which does not take place, until the young fishes have attained some size and have been swimming in the ocean like ordinary fishes. The fins which run all around the flat fish are the dorsal and anal, and the name *Pleuronectidæ* (Greek *pleuron* side, *nektes* swimmer) refers to this peculiarity. In Psettodes we find a sort of transition state in the adult, and the eyes are as often found on the right as on the left side, and not unfrequently they swim in a vertical position. In the adult state they live on the bottom and swim with an undulating motion. They prefer sand, with which they cover themselves. Some enter fresh water, and others never live out of it. One fresh-water species of sole was brought down by me from the Palmer River (*Synaptura selheimi*, Macleay), obtained by Mr. Selheim and others. Günther says that all flat fishes are carnivorous, but this must be subject to some exception, as these fresh-water soles were captured by a bait of grass. The size and abundance of flat fishes and the flavour of the flesh of the majority render this family one of the most useful to man.

There are about thirty-five genera, of which *Hippoglossus*, the "Halibut," *Rhombus*, "Turbot," *Solea*, the "Sole," and *Pleuronectes*, the Plaices, are the most familiar.

## The Flounder.

(Plate XXXIII.)

"Out of the eight or nine species of flat fishes found in Eastern Australia, two only are of sufficient size and frequency of occurrence to be classed among our useful fishes. These are *Pseudorhombus russellii*, generally called the "flounder," and *Synaptura nigra*, best known as the "sole." The first of these is to be found on all the sandy bottoms both inside and outside the bays and inlets of the coast. It

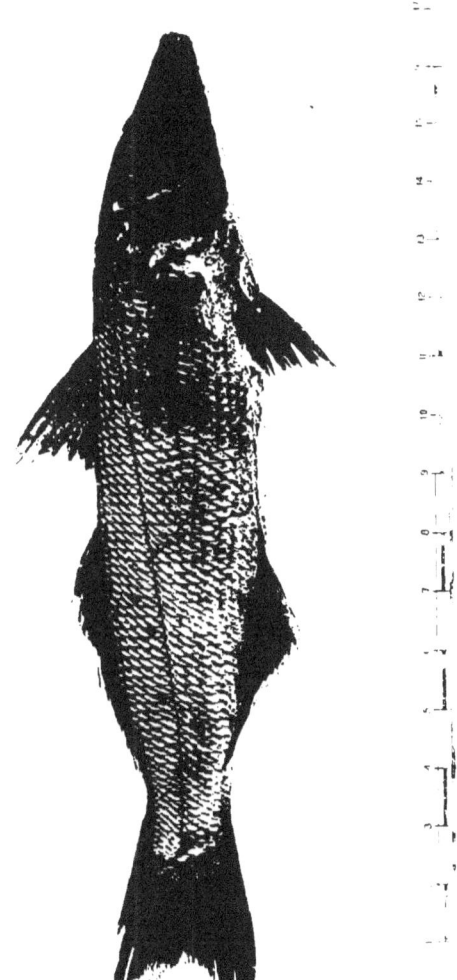

*Sillago ciliata.*—CUV. & VAL., not GÜNTHER

NORTHERN WHITING RARE IN N. S. WALES.

is generally taken by the hook, the closeness with which it adheres to the ground making it merely a chance taking it in the seine. It is a good fish, by many people esteemed considerably above its true merits. The other species, the "sole," is an exquisitely delicious fish, equal if not superior to the Thames sole, but curiously enough is as much undervalued by the public generally as the flounder is the reverse. It is probable that this prejudice proceeds from bad cooking, as the "chair" is of a very tough gelatinous character. The mode of capture is solely by spearing, as it is a fish which never takes the hook, and lies far too close to the ground for the seine. The *modus operandi* is as follows :—On a calm clear morning pass slowly in a boat over the shallow flats where they abound. The fish, alarmed by the approach of the boat, seeks to hide itself in the mud. A small cloud is thus raised, and into the middle of this cloud the fishermen plunges his spear, and unfailingly secures his fish. Large numbers have been taken in this way in one morning at the mouth of Cook's River, in Botany Bay. That and Lake Macquarie seem to be favourite haunts of this fish. Both species come to the flats inshore after midsummer, probably only from deeper flats and banks outside. They are then full of roe, and there can be little doubt that the spawn is deposited on the sandy flats on which the fishes are then found. The trawl net seems to be the most likely way of securing a regular supply of these most valuable fishes.—R.R.C.

Some of the European soles spawn in the open sea, according to Mr. Holdsworth.*

*Pseudorhombus* has a wide mouth, and teeth in both jaws of unequal size in a single series ; none on the vomerine and palatine bones ; eyes on left side, interorbital space not concave, dorsal fin commences on the snout ; scales, small ; lateral line with a strong anterior curve. This is generally a tropical genus. One species, which has been found in North Australia is yellowish brown with two or three spots on the lateral line. There is another species in Port Jackson, *P. multimaculatus*, Günth., which is a greyish-brown with smaller and larger sub-ocellated blackish-brown spots ; fins, finely spotted with brown, a series of larger distinct spots along the basal half of the dorsal and anal fins.

We have also many other genera and species in Victoria. The sole is *Rhombosolea bassensis*, Cast., and the Flounder, *R. plesoides*, Günth. We have also in Port Jackson two species of the true sole, viz.,—*Solea microcephala*, Günth., and *S. macleayana*, Ramsay. The genus *Synaptura* is peculiar to the Indian and Australian seas. It has the eyes on the right side, upper in advance of the lower, cleft of mouth narrow and twisted to the left side. Teeth minute, on blind side only, none on vomer or palate. Vertical fins confluent, scales small, ctenoid. Lateral line straight. We have three species, and there are two others in the tropics, one of which is the fresh-water species already referred to.

## The Sole.

(Plate XXXIV.)

*S. nigra*, Macleay, is black on the right side and the left yellowish white. The vertical fins are tipped with white. The scales are firm and hard, and their apices glassy-looking, with seven or eight acute points, those on the blind side like them but not so strongly armed. In the harbour it never attains a length of more than 10 inches, and about 6 wide (high). Mr. Macleay having tried some experiments with a trawl net obtained some of much larger size. He regards the fish as superior to the English sole.

* See article " Fisheries " in the ninth edition of the Encyclop. Brit.

The following is the experience of Mr. E. Hill in capturing these fish :—"The flounder and sole are the only representatives, so far as we know, in our Australian waters. It was at one time alleged by a French officer visiting this country that he had caught a turbot while off Sydney Heads. Some years subsequent the late Sir W. Denison and Sir Daniel Cooper fitted out a trawling expedition along the coast. A vessel of over 100 tons was placed at their disposal by Messrs. Broomfield and Whittaker, of this city, manned by the ordinary crew, together with a number of additional hands placed on board by Captain Denham, of H.M. ship "Herald," under charge of Captain Hixson, our present Superintendent of Pilots, who, together with the writer and Captain Broomfield, proceeded to the scenes of operation, which extended from Broken Bay on the north to Jervis Bay on the south, and which voyage lasted over twenty-eight days, without any result for the fish sought, although thousands of fishes were caught during the time, including the flounder and the sole, but no turbot.

"The flounder is not very abundant, and is caught by line and hook as well as by net. With line and hook they have been frequently caught over towards Manly Beach, and on the Sow and Pigs sandspit, as well as at other places in the harbour. Botany Bay also affords some particular spots where they are most likely to be; but there is little or no certainty of catching a flounder at any time—they come in promiscuously.

"The sole is one of those rare fishes of this Colony that the flounder is very frequently made to take its place. 'Pay your money and have your choice.' Call it either flounder or sole, as you please. There are few who know the distinguishing features between flounder and sole, but many recognize the distinction when they have them together.

"The flounder takes a bait as he is swimming, and is armed with rather formidable teeth set in a mouth across. The sole will not take a bait, nor does he feed except only when perfectly flat on the ground, dark side up, and generally with sand and *debris* to the eyes, leaving his mouth, which is formed very like the letter S and without perceptible teeth, free to take in food.

"The usual or only method of taking the sole is by spear during the calm mornings of winter when the water is clear; the slightest ripple is a serious obstacle. This is sometimes overcome by a little grease or oil on the surface of the water, and it is usual to take a piece of fat meat for that purpose. The spear should have but one fine steel prong, with which you probe the sand on their feeding patches, and when the fish is pierced it makes no resistance, and is easily brought to the surface. Very frequently two are pierced and brought up at the same time, and in one or two instances I have seen three brought in with one probe by striking the spot where they had accidentally over-lain each other in the feeding patch.

" The sole of this country is a very delicious fish when properly cooked. No dependence, however, is to be placed in the certainty of a catch, and I know of no grounds more likely than that of Cook's River channel above the Waterworks, or the Wallanora, up George's River; at this latter place I have been successful, having taken in one morning over 100 pair."

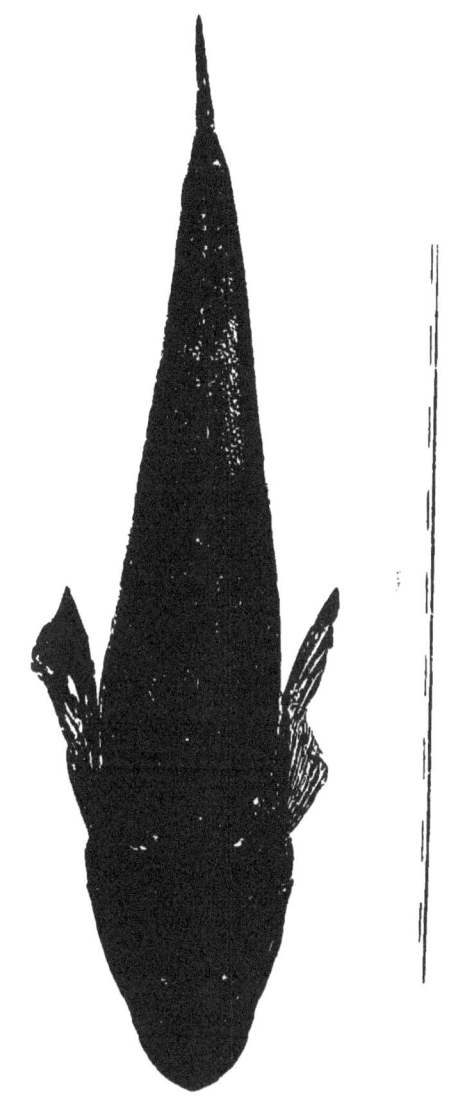

*Platycephalus fuscus*, Cuv. & Val.
FLATHEAD.

Mr. Lee Lord, in a paper sent to the Royal Commission, gives an equally interesting account of this fish. He says:—"I procure it in large numbers at the mouth of Cook's River, and also near the mouth of Shea's Creek, where the water is shallow. The depth varies from 4 feet to 12 feet, but averages about 6 feet. My mode of capture is with the net or spear, for they never take the hook. I use a spear 12 feet long, with one steel spike at the end 9 inches long, without a barb and quite smooth, as in this way the fish is not torn or injured. The water requires to be as smooth as glass, so that the bottom or sand may be easily seen, and the early morning is the best time. Low tide is most suitable, for the boat is then more easily managed and the water not so deep. When the tide is not too strong I propel the boat with the spear, so as to follow along the banks and holes where the fish lie; if a strong tide, allow the boat to drift. They lie in the sand, mostly covered all over, but some only partially, when they are very easily seen. As they feed the sand is disturbed, and I find them either in holes about 4 feet square and 6 inches deep, or else by following their tracks along the banks. By the latter method I have often gone 15 yards before finding the fish. The first method is the most common, and in some of these holes I have taken four fish (for if you spear one, the other will not move till touched). In the summer months they usually feed along the edge of the channel; in winter they go into a little deeper water, and I find them then generally just inside the ledges or banks of sand formed by the tide; they are more difficult to find.

"On the 21st February last I speared $2\frac{1}{2}$ dozen in less than $1\frac{1}{2}$ hour, the best lot I ever procured, nearly all the same size, very fat, and full of roe. Those who tasted them pronounced them the most excellent they had ever seen. In January, February, March, April, October, and November I have always got large numbers,—as many as five dozen before breakfast,—and in these months I have found them most plentiful, sometimes in one part sometimes in another, say fully 1 mile apart. In the winter months, May, June, and July, I have taken large numbers, averaging $1\frac{1}{2}$ to 2 dozen per day, but the fish is not then such good eating.

"The appearance of the sole is oval-shaped, with little or no tail, *grey* colour, very much the colour of the bottom; many are spotted all over the top side, and get darker after being out of the water some time. The bottom side is quite white; the eye is very prominent, and often looks like a pearl on the sand, and is sometimes the only method of distinguishing the fish.

"To prepare for cooking they require to be placed in boiling water from ten to fifteen seconds, when the skin begins to come away from the outer edge, and is then easily drawn off top and bottom, with a dry cloth, from the tail to the head. The skin will not come off without scalding water."

The Hon. W. Macleay, who has always taken a very active interest in our fishes, has held the opinion that the *Pleuronectidæ* are not less frequent in our own seas than any others. As however they are all ground fishes and except in the spawning season always keep in deep water, and seldom take a bait, the only way to capture them is by the use of the trawl net. The experiment had not been previously tried except in a clumsy way, which was not successful.* The Government of New South Wales, acting on the recommendation of the Royal Commission on the Fisheries of New South Wales, lately imported a variety of nets, lines, and fishing implements of the latest and most improved kinds, from England, Norway, and America. Amongst these there were two trawl nets, a large Grimsby beam trawl and an otter trawl of 42 feet width of net. Early in January of this year (1880) a trial of the other trawl was made. The net was put overboard off Middle Head, and was raised in North Harbour, near Manly Beach. It was again lowered at the mouth of Middle Harbour and raised opposite Clontarf. On both occasions the net was quite full. Besides a number of rays and sharks, there were gurnards, gobies, and a large number of *Pleuronectidæ*. The flounder was got in considerable number, though rarely taken in the seine net. (Mr. Macleay thinks that of all European flatfish it nearest approaches is the "Brill," *Rhombus lævis*). Other rare flatfishes were found, including a new species of *Synaptura* and a new genus entitled *Lophorhombus cristatus*. One very large sole and several smaller ones were obtained. So large a capture where such a small space was traversed by the trawl, and in not very deep water, shows what advantages may be expected if trawl-fishing becomes an industry in New South Wales. It was singular, Mr. Macleay remarks, that no species of *Rhombosolea* were found, though represented by many species in Victoria, Tasmania, and the south coasts generally. He predicts moreover that a better acquaintance with the deep-sea fauna, by the aid of the trawl net, will prove the existence in the deep-sea currents of species of *Rhombosolea* like the *R. monopus* (New Zealand), rivalling the European turbot in size and excellence.

## IV.—Ord. PHYSOSTOMI.

All the fin-rays articulated, first dorsal and pectoral sometimes ossified. Ventrals, if present, abdominal, without spines; air-bladder, if present, with a pneumatic duct (except in SCOMBRESOCIDÆ).

## SILURIDÆ.

No scales, sometimes osseous scutes, barbels always present; maxillary bone rudimentary, almost always supporting a maxillary barbel. Margin of upper jaw formed by the inter-maxillaries. Subopereulum absent. Air-bladder present, generally communicating with the organ of hearing by ossicles, an adipose fin, or none.

There are several species of *Siluridæ* in these seas, the most common in Port Jackson being the "cat-fish" (*Cnidoglanis megastoma*, Richardson). Though an excellent fish it is very seldom eaten; the prejudice existing against the whole family being almost universal among Europeans. It is frequently taken in the net on the muddy beaches of the harbour.—R.R.C.

* The substance of these remarks is contained in a paper read before the Linnean Society of New South Wales, in January, 1882, and published in the 7th volume of the Proceedings.

## The Cat-fish.

Mr. Hill says that they are chiefly nocturnal, or in day-time lie about in secluded spots or on the edge of the fringe weeds. Cat-fish are very numerous at several places, and when out net-fishing they are very troublesome. The head is armed with two or three heavy spines, placed in such a way that they are most effective—one bone on each side of the head, low down and far back, and another at the commencement of the dorsal spine, nearly at the back of the head. These sharp bony substances are concealed by a thin membranaceous film the colour of the fish, but which are easily bared, when they appear white. There can be no doubt that these spines are really venomous, and to be punctured by one of them is a serious matter. Not only is the pain intense, but the after consequences are generally grave. All the Siluridæ have these spines as weapons of defence, which can be secured by another spine, which acts as a bolt or fulcrum, and a bony ring, along which some sort of virus seems to be secreted. This marine cat-fish is quite black. They are distinguished from other freshwater cat-fish (*Copidoglanis*) by having all the gill membranes united below the throat, and attached to the isthmus along the entire mesial line. The mouth is overhung with fleshy cirrus, and this together with their being nocturnal causes them to bear the name of cat-fish.

"The cat-fishes make a hole or mound, to which the female may retire at the summer season to incubate. I have often seen the blacks get them out with very primitive means. First they procure a straight thin limb or piece of scrub tea-tree or other wood, about the size of a stout ramrod and twice as long. This they point and harden in the fire, and then with another piece of stouter wood or a dingy paddle they proceed to work, two or three stabbing in the same mound or patch, for it is raised only a very few inches, till one or another has pierced the old one ; then with the short stout piece of wood or paddle they dig it out, and very often as a preliminary they will examine the spot with their foot, but are very careful then, and can readily detect if one is at home. The blacks appear to be very fond of these fishes ; and at this season of the year, while the fish is roasting, and the rich-looking ovaries are bursting through the heat, they look very tempting ; and I think this is the best way to cook them and the proper time to eat them. They give to fishermen a great deal of trouble with their nets, and often get their thorns entangled endeavouring to get through the meshes. William Anan, an aboriginal, accompanied a party net-fishing to the head of Darling Harbour one afternoon ; the place was already celebrated for cat-fish, and after the net had been hauled, and being prepared again for the boat, a part of poor William's hand came in contact with one of the thorns of the cat-fish. This gave him great pain, and soon commenced to inflame, in consequence of which he had his arm amputated to save his life. He lived long afterwards, and was useful and cheerful when out with fishing parties."—E. S. H.

The other species of cat-fish *Copidoglanis tandanus*, Mitchell, will be dealt with amongst the fresh-water fishes.

L

## Fam. SCOPELIDÆ.

Like Siluridæ, but body often scaly; no barbels, no air-bladder. Opercular apparatus sometimes rudimentary, an adipose fin. Pyloric appendages few or absent. Intestinal tract very short. Exclusively marine forms.

"They are all deep-sea fishes, and excellent for food; but of the five species known in these seas, one only, the "Sergeant Baker" (*Aulopus purpurissatus*), is of any size. It is a beautiful as well as a good fish, and is frequently caught by the hook in the summer season by the schnapper-fishers. The other species of our waters belong to the genera *Saurus* and *Saurida*."—R.R.C.

The genus here referred to has the head and body rather elongate, slightly compressed, covered with scales of moderate size. Mouth very wide. Maxillary well developed, dilated behind. Teeth small, heart-shaped, in bands in the jaws, on the vomer, palatine, and tongue; eye moderate. Pectorals and ventrals well developed, the latter nine-rayed, inserted close behind the pectorals below the anterior dorsal rays. Dorsal in the middle of the body, rather long, with fifteen or more rays. Adipose fin small, anal moderate, caudal forked. Gill opening very wide.

## The Sergeant Baker.

(Plate XXXV.)

This fish is *Aulopus purpurissatus*, of Richardson, is a species distinguished by having the second and third dorsal ray produced into a long ray in the males. The colours of this fish are very brilliant, consisting chiefly of purple and red. Mr. Hill says of this fish:—"The red gurnard or gurnet, popularly known as 'Sergeant Baker,' is of the genus *Aulopus*, 'combining the character of the salmon and the cod-fish' (Cuvier).* It was long supposed in this country that the red gurnard or Sergeant Baker and the flying gurnard were of the same family and genus, the distinction being merely in the wings; the colour and similarity in shape no doubt was the cause of the delusion. The Sergeant Baker in all probability got its local appellation in the early history of the Colony (New South Wales), as it was called after a sergeant of that name in one of the first detachments of a regiment, so were also two fruits of the Geebong tribe (*Persoonia*); one was called Major Buller, and the other Major Groce, and this latter again further corrupted into Major Grocer; such was the spirit of corruption in those days.

"It is not only a good fish, but combines the flavour of the salmon with the rich flakiness of the cod-fish. It can readily be understood why connoisseurs select them, after a day's sport with line or net, for they may be caught with either, and at any place, but not many are secured at the same time, and these fishes are what may be termed scarce. I have known the Sergeant Baker and the flying gurnard caught with hook and line out of the same boat and off the same ground; both had their heads cut off and fried in the same pan, but there was a positive distinction, although both were good."

* There is nothing of the salmon about it, except the adipose fin.

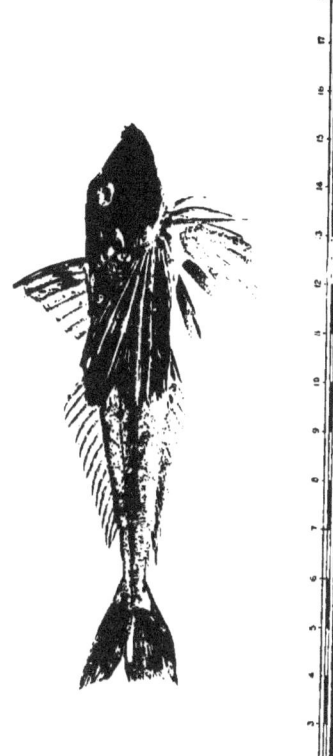

*Trigla polyommata.*—RICHARDSON.

THE FLYING GURNARD.

It will not be necessary to delay over the next eight families of fishes any more than to state that they include the immense groups of *Cyprinidæ*, Carps, which is not represented in Australia, though the Crucian Carp, *Carassius carassius* has been introduced into very many of our rivers and ponds. There are none of the Family of *Cyprinodontæ* in Australia though represented by so many genera in Africa, Asia, America, and Southern Europe.

## SCOMBRESOCIDÆ.

Body covered with scales, with a keeled series on each side of the belly. Margin of the upper jaw formed by the inter-maxillaries mesially and by the maxillaries, and by the inter-maxillaries laterally. Lower pharyngeals united into a single bone, dorsal fin opposite the anal and near the tail. Stomach not distinct from the intestine, which is quite straight, without appendages.

Fishes of this family are generally distinguished by some peculiarity in the jaws. In *Belone* both jaws are produced into a long slender beak, no finlets. In *Scombresox* the jaws are the same, but there are a number of finlets between the dorsal, anal, and caudal.

In *Hemirhamphus* the lower jaw only is prolonged into a long slender beak. In *Arrhamphus* the upper jaw forms a triangular more or less convex plate, and the lower jaw is not prolonged. In *Exocœtus* the jaws are short and the pectorals converted into long organs of flying. All but *Arrhamphus* are found on the coasts of N. S. Wales, and are excellent table fishes.

## The Long Tom.

(Plate XXXVI.)

*Belone ferox* is the "Long Tom" of the Sydney fishermen, a "Garfish" which is only taken in the net. They have teeth widely set apart in their long jaws, and they skim over the surface of the water seizing the small prey as birds do with the beak, and which without the teeth they would be unable to hold. They can only swallow small fish, and swim with an undulatory motion of the body. Dr. Günther says that shoals of them appear upon the coast with mackerel. In consequence of their bones being of a green colour an opinion exists that they are poisonous. There are one or two doubtful cases of fish-poisoning on record, said to have arisen from eating the bones.

Of the "garfishes" we have four species known to be found on our coasts. One, *Hemirhamphus regularis*, is the favourite breakfast fish of the citizens of Sydney. *Hemirhamphus melanochir*, or "river garfish," is a still better fish, but has become very scarce. *Hemirhamphus argenteus*, the common Brisbane species, but rare in Sydney, and *Hemirhamphus commersoni*, also rare in Sydney, but abundant in the far north. This last is the largest of the genus, but scarcely equal in quality to the others. The ordinary Sydney garfish (*H. regularis*) comes in from the sea in the latter end of summer, to deposit its spawn in suitable spots in the harbour. It is then in the finest condition, and makes its appearance in successive shoals, some of them of enormous size. During the latter end of February of this year the shoals were so large that several boat-loads were taken in one haul of the seine, and the fish on the spot could be purchased for sixpence a bushel.—R.R.C.

## The Garfish.
### (Plates XXXVII and XXXVIII.)

*Hemirhamphus intermedius,* Cant. (Plate XXXVII upper fig.) has a dark greenish back, with a well defined silvery band on the upper sides. *H. regularis,* Günth. (Plate XXVII lower fig.) (the river garfish of the fishermen). The two species may be readily distinguished by the small triangular upper jaw being in one (*intermedius*) longer than broad, and in the other (*regularis*) broader than long. In *H. intermedius* also the scales are so very deciduous that the fishes generally appear to be without any. *H. commersoni,* Cuv., has four rounded blackish blotches on the sides. *H. argenteus,* Benn., has a very short beak. According to Mr. Hill, " the garfish, or among professional fishermen ballahoo, is another of the delicate fishes of this Colony, and one which is known and recognized for its good qualities by every one who has lived in the city of Sydney, or throughout the length and breadth of the country. These fishes come into the harbour at various stages of growth, and may be found from the size of sail-needles to adults of 15 or 18 inches long. January and February are the months when they may be seen plentiful of mixed sizes, but by a wise enactment of the law nets only of a particular gauge in the meshes are allowed to be fished with, to prevent the wholesale destruction which used to go on in former years, when millions of the smaller and useless garfishes were hauled on to the shore to perish.

" On the 1st of April, however, that restriction is taken off, at which time nets with meshes of a suitable size are substituted, and we may then expect a fair supply to the market, as our harbours are now teeming with them, and the season extends into the winter months."

None of the true Flying Fishes are caught near Port Jackson, but they may be seen in summer months on the more northerly portions of the coasts of New South Wales.

A great many families must now be passed over, as they include none of commercial importance to Australia. There are no true *Salmonidæ* known in any of our waters, but Salmon and Salmon Trout have been successfully introduced. An attempt has been made in Tasmania to acclimatize *Salmo purpuratus,* as well as *Salmo salar, S. fario,* the Trout, and *S. trutta,* the Bull Trout or Salmon Trout. So far the experiment may be said to be pending. Whether *S. salar* is really acclimatized or not is still doubted by some, though for no good reason, the largest specimens, over 10 lbs. which I have seen caught were certainly true Salmon, according to the opinion of the most experienced persons in Tasmania. The whole evidence on this question will be stated further on in connection with acclimatization of Fishes.

## CLUPEIDÆ, or Herrings.

Body covered with scales, head naked, no barbels, abdomen frequently compressed into a serrated edge. Margin of upper jaw formed by the intermaxillaries mesially, and maxillaries laterally, the latter composed of at least three movable pieces. Opercular apparatus complete, no adipose fin, dorsal moderate, anal sometimes very long. Stomach with

PLATE XXVIII.

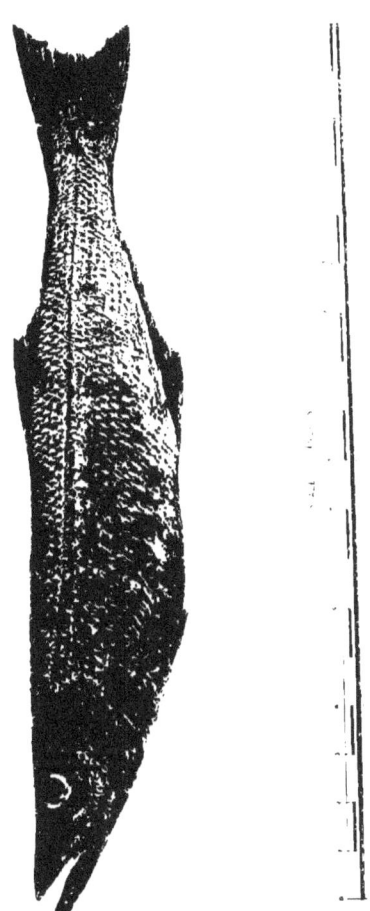

*Sphyræna obtusata.*—Cuv. & Val.

THE PIKE OF PORT JACKSON.

a blind sac, pyloric appendages numerous. Gill apparatus much developed, the openings generally very wide. Air-bladder more or less simple. Dentition feeble.*

This family is not numerous in species, but is the most numerous in individuals known. It comprises coast fishes spread all over tropical and temperate seas. It was at one time thought that we had no true Herrings, or if present not as numerously as the northern representatives of the family, or as useful. The Hon. W. Macleay has sufficiently disproved this, from whose essay on the subject† large use is made in the following remarks.

Herrings are rarely seen in our markets, but this is due to the fact that the shoals do not as a rule enter our harbours, and to fish for them in the open sea requires appliances not at present in the possession of our fishermen. Our species are different from those of the northern hemisphere, but in excellence as food they are not inferior.

Anchovies or *Engraulis* have a compressed body with a very wide lateral mouth, and a projecting upper jaw. Scales large. We have two species, *E. antarcticus*, Casteln. and *E. nasutus*, Casteln. The first-named species is by many erroneously believed to be identical or at most a variety of *E. encrassicholus*, of Europe. Count Castelnau states that it is very common in the Melbourne market at all seasons, and goes by the name of "Whitebait." Mr. Macleay says it is never seen in Sydney markets, or any species of *Engraulis*. He thinks that from its being common at all seasons that it must be inferior to the real Anchovy at least in economic value, because it belongs to that class of fishes whose visits elsewhere are periodic and in enormous quantities; that as this anchovy does not occur in shoals, its fishery could never be of much value. He however also considers it not unlikely that the specimens may be young fish, and that the periodical haunts of the large shoals have not yet been noticed.

*Chatoëssus.*—Mouth transverse, inferior, narrow, without teeth, upper jaw overlapping the lower, abdomen serrated. Two species, *C. erebi*, *C. richardsoni*. These are fresh-water fishes which will be dealt with separately. *C. erebi* is sometimes caught on the coast, and is called "Bony Bream."

*Brisbania.*—Mouth wide, opening upwards, maxillary large, teeth small and numerous, last ray of dorsal elongate, abdomen not serrated. There is only one species of this fish, which is of a genus erected by Count Castelnau for one known as yet only in the Brisbane River.

*Clupea.*—True Herrings. Mouth small, teeth minute or none, abdomen serrated, anal fin short. We have ten species if we include the true sprat (*C. sprattus*) which is said by Günther to be found in

---

* In 1873, the number of herrings brought into Yarmouth and Lowestoft (England), was 423 millions, of a value of £875,000. In France, in the same year, the value of the herring fishery was £400,000; In 1876, the value of the same fishery was £825,620. In Ireland, in the same year, it was £226,803. In Scotland it has been yearly increasing, and in 1876 reached 598,197 crans.

† On the *Clupeidæ* of Australia. Linn. Soc. N.S.W., vol. 4, p. 363.

Australia. *C. sagax, C. sundaica, C. hypelosoma, C. moluccensis, C. tembang. C. Novæ-hollandiæ, C. vittata, C. richmondia, C. schlegelii,* and *C. sprattus*, are the Australian representatives.*

## The Maray.

*C. sagax*, Jenyns, is almost identical with the English pilchard. Prof. McCoy states that a specimen was brought to him in August, 1864, from a small shoal then seen for the first time in Hobson's Bay. In the same month of the year succeeding they appeared in great abundance in the bay, and were caught by thousands for the market. After remaining for a few weeks they disappeared until the same time in 1866, when they appeared in such numbers that baskets could be easily filled by simply dipping them into the sea. Hundreds of tons were sent up the country to the inland markets, and through the city for several weeks they were sold for a few pence the bucketful. It was known in Port Jackson many years prior to the date above-mentioned. It may visit the coast annually, but only by chance enters the harbours. It is called the "maray" in New South Wales, probably a native name.

The "maray" (*Clupea sagax*) appears annually in immense shoals about midwinter, passing in a northerly direction, and portions of the shoals sometimes enter the bays and harbours of the coast, but certainly not, as with most fishes, for the purpose of spawning, for the shoals consist at that time of small and immature fish, most probably driven in by the hosts of large fishes, porpoises, &c., by which they are invariably pursued. The same fish is seen to pass south on the eastern coast of New Zealand about six or seven months afterwards, and then they are fullgrown and full of roe. The excellent bloaters of Picton, New Zealand, are made of this fish.—R.R.C.

The shoals are described as enormous, covering miles of sea, and accompanied by flights of birds and numbers of large fishes. They are generally observed from 1 to 3 miles from the land, followed by multitudes of gulls, mutton-birds in the air, and barracoota, porpoises, &c., in the water. It would be very interesting to inquire how far north they go, and what are their spawning grounds; facts unknown to us hitherto.† It is unknown in the warm seas of North Australia, or the Indian Archipelago, yet it is found in Japan and California, and other temperate regions north of the equator.

## The Southern Herring.

The next best known species is the "herring" (*Clupea sundaica*), a fish of great excellence and delicacy of flavour, though but little appreciated. It also appears in the winter season and in shoals. It is said

---

* If it be true that the sprat is only the young of the common British herring, which is believed by some good authorities, then we certainly have not the species amongst our fishes. The difference between the herring and sprat is that of size, and the possession of vomerine teeth by the former, as well as other distinctions such as the shape, formula of the fins, the smell, the taste, the position of the fins, and the presence of a black line on the tail. See "Buckland's British Fishes," p. 241.

† It is thought that the herring always spawns in shallow grounds, because the ova are found on shallow bottoms, in a glutinous mass, mingled with clay, sand, weeds, &c. Mr. Holdsworth however points out that the specific gravity of the ova causes them to sink, and also that the roe is known to be shed at sea. See article "Fisheries" on this subject, Encyclo. Brit., ninth edition, where also the interesting observations of Prof. Allman on the matter are given.

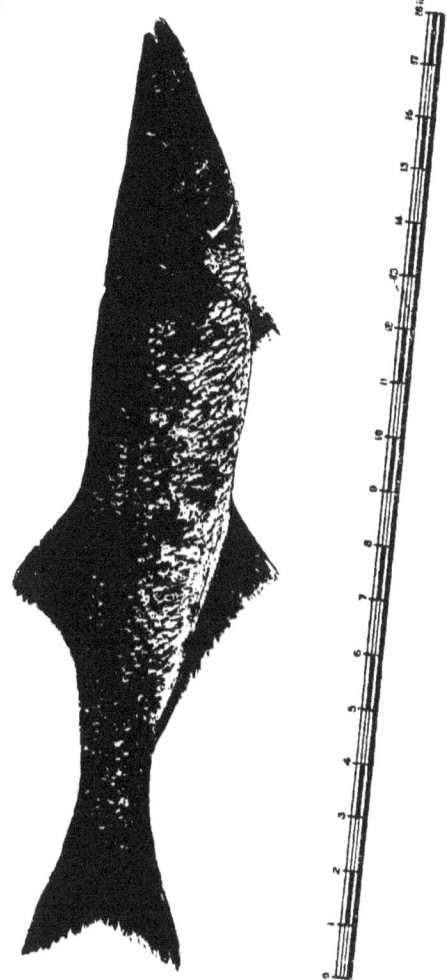

to be found about the mouth of the Hawkesbury River at all seasons of the year. It can easily be distinguished from *C. sagax*, by its much deeper and more compressed body, its deciduous scales, and the bright golden bands on each side of its back. It is about 7 inches long. It is as well known to the fishermen as the herring, and visits the coast like the last in enormous shoals which travel in a northerly direction. It seems probable that its breeding-ground is not far distant, as some are found in the Hawkesbury at all seasons, and the young fry, of apparently the same species, sometimes abundant. The adult fish is caught in great numbers in Java and Celebes, and forms a very important article of food to the population. In excellence and delicacy of flavour, Mr. Macleay thinks them superior to the common herring of Scotland, and that preserved in oil like sardines they would rival those delicacies.

The same herring is rather common in the Upper Hunter at all seasons, but it is small. At West Maitland the anglers value the fish for sport, though it very seldom exceeds seven inches in length. But it is delicious eating. It is best caught with the common house-fly dropped gently upon the water and moved along the surface with caution.

*C. hypelosoma*, Bleek, is very like the last ; it is proportionately deeper, the maxillary bones seem to extend further back under the eyes, and there is no golden band along the upper part of each side. It is not unfrequently seen in Port Jackson, and is called herring also. It is considered equal in flavour to the " Maray." It is said to pass the Sydney heads in enormous shoals like the last, and to mingle with them.

*C. mollucensis*, Bleek, is a Molluca fish said to be seen in Port Jackson ; if so, it must be extremely rare. Nothing is known of its character.

*C. tembang*, Bleek. North Australia—rare.

*C. Novæ-hollandiæ* Cuv. and Val. This and the two following species are the Australian representatives of the sprat ; they are mostly fresh-water fishes. The one named is a beautiful little fish about 5 inches long, known as the herring in all the tributaries of the Hawkesbury, perhaps not in any other eastern rivers, and certainly not in any of the western ones. Angling for this fish is a favourite sport in some of the upper waters of the Nepean, but it is of no value as an article of food.

*C. vittata*, Castelnau. Abundant in the Melbourne market at times, and known as the smelt. It is probably a fresh-water species.

*C. richmondia*, Macleay. A fresh-water species peculiar to the Richmond River, which much resembles *C. novæ-hollandiæ*, except that it is smaller and has a broad silvery stripe on the sides, margined above and below by a dark stripe.

*C. schlegelii*, Castelnau, is a rare species from Port Darwin, N.A.

*Sprattelloides*. Mouth anterior and lateral, abdomen not keeled, dorsal fin opposite ventrals, no teeth. Only one species, *S. delicatulus*, Benn. It was seen in enormous shoals at Darnley Island during the fortnight that the " Chevert " lay there. At that time, the early part of August, 1875, the whole northern shore was literally black with them, and there would have been no difficulty with proper appliances in preserving hundreds of tons of these finest of all sardines (Macleay, Proc. Linn. Soc. N.SW., 1st vol. p. 351).

*Etrumeus jacksoniensis*, Macleay. Distinguished by the absence of compression or serration on the abdomen, and the position of the ventral fin quite behind the dorsal. Only one species is known. It is also called Maray, and is said to pass northwards in enormous shoals every winter, like all the other herrings of our coasts. It is very good eating.

*Elops*. Upper jaw shorter than the lower, abdomen rounded, an osseous throat-plate, scales small. A very beautiful fish, rarely taken in Port Jackson, as it is strictly speaking tropical. It is not much in request for food.

*Megalops*, distinguished by its larger scales. One species, *M. cyprinoides*, about a foot long, is found in the Hawkesbury, but has a very large range. Dr. Cantor says of this fish, or of one very closely allied, that notwithstanding the numerous fine bones, the species is valued for the flavour, and is multiplied rapidly in tanks and fattened for use. " Fishes of Malacca," p. 289.

*Chanos*. Mouth small, toothless, abdomen flat, gill membranes entirely united, scales small.

*C. salmoneus*. A uniformly silvery fish, which is the most highly prized of all the herring family in consequence of its excellent flavour. It is reared in tanks in India, and reaches a length of about 2 feet. "If a little of the enterprise exhibited in the efforts that have been made to introduce the salmon into our rivers were expended upon the cultivation of this fish, in our coast rivers north of the Clarence, the result, I venture to say, would be more satisfactory." Macleay.

## EEL FAMILY, MURÆNIDÆ.

Body elongate, cylindrical or band-shaped, naked or with rudimentary scales; vent a great distance from head, no ventral fins, vertical fins absent or confluent or separated by the projecting tip of the tail. Sides of the upper jaw formed by the tooth-bearing maxillaries, the fore part by the intermaxillary, which is more or less coalescent with the vomer and ethmoid. Humeral arch not attached to the skull; stomach with a blind sac, no appendages. Spread over almost all fresh waters and seas of the tropics and temperate zones, some descending to great depths. In the majority the branchial openings are wide slits, in the true *Murænæ* they are narrow.—G.S.F.

## The Eel.

(Plate XXXIX.)

Eels are caught in tolerable abundance at times on all parts of the coast, but their capture is only accidental when seeking for other fish—such a thing as fishing for eels being unknown to our fishermen. There are four specimens of good size and quality. The common eel (*Anguilla australis*), really a fresh-water fish, but descending to the sea at certain periods to spawn. The "sea eel," or "silver eel" (*Murænesox bagio*), a very fine fish, and never common. The air-bladder in this species is of very great length, and is probably of value for isinglass. The

*Mugil grandis*—CASTELNAU.

THE SYDNEY SEA MULLET.

"conger eel" (*Conger labiata*) also grows to a large size, but it is rarely brought to market. The "green eel" (*Murœna afra*) is found abundantly in the holes and crevices of the rocks everywhere.—R.R.C.

A figure is given of the Australian eel. The silver eel is not uncommon in the Hunter River and is caught by night-lines just in the same manner as the ordinary freshwater kinds. The other genera and species need no further remarks than those given above.

It is a singular fact in the distribution of our fishes that eels in Victoria are only found in rivers which take their rise on the south side of the dividing range. It is also asserted that no eels are found on any of the western waters, but this is doubtful. It is quite certain that the Murray cod *(Oligorus macquariensis)* is found on both sides of the dividing range. On this subject Mr Hill says:—"So far as the cod-fish is concerned, I have seen it caught in the Upper Clarence, where many eels exist, and some two or three hundred miles (as the river goes inland) towards the head of eastern waters. I have seen and partaken of remarkably fine cod, identical with those of the Murray, and which I have also caught in the Murrumbidgee. Mr. Wilcox, of the Clarence River, I think, first called attention to the fact that these cod-fish did exist in the eastern as well as in the western waters, and since which it has been asserted that eels have been taken in particular parts of the western shed. The report that eels are said to have been caught in the western shed will probably cause, now that attention has been called to the subject, many cases wherewith this may be corroborated and multiplied, or it may appear that they have been only the *Plotosus* or eel-fish, described by Sir Thomas Mitchell." The fish here referred to is *P. anguillaris*, Lacepede, one of the Siluridæ, or cat-fishes.

To fish for eels it requires a well secured bait and night lines; this has not been much tried in the proper places of the western waters, it having been taken for granted that the hypothesis referred to has been correct; certain, however, it is the fact, that these fishes have not been found to be plentiful or well disseminated, either the eel in the western or the cod-fish in eastern waters. There is no reason why eels should not be in congenial places of the western shed; but that which appears to me to militate against their general dissemination in these regions is great altitude and cold water, racy bottoms and rapid streams, and further west the liability to continued droughts.

## The Leather-jacket.

(Plate XL.)

The orders *Lophobranchii* and *Plectognathi*, containing the curiously formed fishes known as "sea-horses," "cow-fish," "toados," "porcupine fishes," "sunfish," and "leather-jackets," are very numerously represented in the Australian seas, but out of the entire number one only can be cited as being in the least degree useful, and that one is productive of more harm than good. It is a "leather-jacket" (*Monacanthus ayraudi*). It is said, when skinned, to be excellent food, but it is a most serious annoyance to the fishermen, infesting their favourite fishing-grounds and cutting their lines. The plague of these fish seems to be on the increase, and unless some means can be found of getting rid of the pest, schnapper-fishing will have to be conducted with wire lines.—R.R.C.

The species referred to belongs to a genus of which we have no less than thirty-seven species in Australian waters. It belongs to the family

SCLERODERMI, of the order Plectognathi, in which the jaws are armed with a small number of distinct teeth, and the skin has shields or is merely rough. In *M. ayraudi* the skin is rough but velvety, the colour is brownish, with three or four whitish longitudinal bands. It attains a length of 18 inches.

It will be necessary to say a few words about the toad-fishes and porcupines, because they are so poisonous, and yet abundant about all our harbours. In 1874 a fatal accident occurred to some children at Coogee Bay. They were out on a picnic, and having caught some toad-fish, cooked and ate them. They died in terrible agony a few hours afterwards.

These fishes belong to the family GYMNODONTES and the genus *Tetrodon*, thus described by Günther:—

"Body more or less shortened. The bones of the upper and lower jaw are confluent, forming a beak with a trenchant edge, without teeth, with or without median suture. A soft dorsal, caudal and anal are developed,—approximate. No spinous dorsal. Pectoral fins, no ventrals. Marine fishes of the temperate and tropical regions. Some species confined to fresh water."

Before appending the observations of Mr. Hill on these fishes, it is necessary to remark that he speaks of the liver of some of these species as being edible. No doubt it may be the case for some kinds, but it is equally certain that the liver is the most poisonous part of others. Sir John Richardson, in his article on Ichthyology in the Encyclopædia Britannica, 8th edit., vol. xii, p. 331, gives an instance wherein two sailors were poisoned at the Cape of Good Hope from eating the liver of one fish of the genus *Tetrodon*. A detailed account of the symptoms was drawn up by a surgeon in the Dutch navy who attended the men until they died, which was in less than half an hour after eating the liver. It should also be remarked that, though Mr. Hill identifies the fish which poisoned the children as *T. hamiltoni*, yet this is questionable. But of nineteen known Australian species seven are found in Port Jackson. Some of them reach a length of nearly 2 feet, but such a size is rare.

## The Toad-fish.

"The toad-fishes do not grow to a large size, at all events the largest we have seen measures something over a foot, and when they do attain anything approaching 6 inches in this Colony they are chiefly found along the coast outside; there they may be found to the size of 8 or 10 inches in length. Those toad-fishes in the harbour are in the young stage, and may be found pushing their way and following the flowing tide to its utmost limits.

"The toad-fishes of Port Jackson alone already described number not less than six varieties, though some of them differ but slightly in appearance, and the whole may be recognized by one common form. That which caused the death of those children alluded to, together with a cat which had evidently partaken of the vomit, is of the genus *Tetrodon hamiltoni*.

*Cossyphus gouldii.*—RICHARDSON.
THE BLUE GROPER.

# FISH AND FISHERIES.

"These fishes have often afforded amusement to the youngsters, as they bite greedily, and may be caught with a pin hook attached to a few yards of thread; and when landed they immediately commence to inflate and puff themselves out to their full tension, and when returned to the water they will collapse and assume their regular shape, and return again to the bait.

"This, again, led to a species of cruelty which gave a particular relish to the boys, especially to those who were inclined that way. After the capture of a fish, and it did not immediately commence to inflate itself, it had to undergo the process of being rubbed on a rough stone, when it instantly commenced a sucking noise, and was soon in a spherical form. It was then placed on a flat solid rock or junk of stone, in a prominent position, and another stone hurled down with all the captor's might on the unfortunate fish, and if the aim was sure, the sudden bursting would make a report nearly as loud as a pistol.

"Of their poisonous qualities there can be no doubt, as it was not uncommon at one time for certain storekeepers to employ a few boys to catch these fish at the rate of fourpence per dozen for the purpose of destroying the rats with which their places of business were infested; and these fishes had the desired effect, for the dead rats were produced in scores, and many still remained dead in their holes.

"The porcupine fishes are of the same family as the toads, and are of the genus *Diodon hystrix*, *D. maculatus*, and *Dicotylichthys punctatus*, with probably some others. These fishes are frequently hauled in with the seine, and are at once discarded by the fishermen. There does not appear to be much flesh on their bones, but I have seen the natives eat *the liver* with great gusto, and, curious to remark, whilst at the island of Fatuna, several porcupine fishes were caught with the hook and line, and which were begged by the natives, who instantly made an incision in the side of the already inflated fish and drew out the liver, which they instantly devoured in its raw state, but the body they threw overboard.

"This is almost sufficient proof that even the porcupine's flesh is poisonous also, and too much care cannot be exercised in meddling with these fishes, either as an article of food or leaving them within the reach of the inexperienced. Even the savages of the Line Islands appear to know of their properties, and in times of feudal warfare they will use the fish only in its inflated and dry state as a covering for the head—not to ward off blows, but to make them look brave and fierce in the encounter. Some of the dried fishes may be seen in our Sydney Museum, as also a drawing (published in the Cruise of the American Squadron, under Captain Wilkes) of an islander dressed in this fighting costume."

*Diodon* is only distinguished from *Tetrodon* by having no mesial division in the teeth. It has also very much longer and stronger spines over the body, and when the fish blows itself out the spines are erected into a most formidable defensive armour. The skin is remarkably hard and thick, and for this reason is stretched and dried by the Esquimaux, who make it into an effective helmet.

## CHAPTER IV.

### Sharks.

These unamiable predacious fishes are very well or rather very numerously represented in Australia, as the list to be given will show. They are not, however, wholly useless. It has been mentioned already how abundant some species are in Tasmania, where the fishermen find their captures of king-fish much interfered with by sharks, which snap them off the hooks. This is probably the Tope or School Shark (*Galeus australis*), Macleay. It was generally thought that it was identical with the Tope of Britain (*G. canis*), but Mr. Macleay has pointed out many differences. It grows to about 6 feet long, and has become an article of export from Tasmania. It is regularly captured at Southport and Recherche Bay. At a visit paid by the writer to that locality he found about ten families engaged principally in the shark fishery. The portions used were the liver, which was boiled for oil,* and the fins, which were dried and pressed for the Chinese market. There were two very intelligent Chinese agents, who bought the fins on the spot and undertook the water transport to the place of shipment. They gave much information about the trade, which just then happened to be in a depressed state. The tail fin was not used, and of the others the dorsal fins were the most esteemed, and were packed separately. The fishermen found it sufficiently remunerative to engage in no other fishery, except when the market was very low, as it was then. There was much sickness in the place about that time because of the way in which the decaying carcasses of the fish were left about. I counted over sixty near one dwelling, and the stench was fearful. There were times when the weather was unfavourable for the fishery, and then mutton-fish were speared. This is the ear shell-fish (*Haliotis nævosa*), which was eagerly bought by the Chinese merchants. Only the large muscular sucking disc or foot is used. Before being packed it is boiled and dried. About 9d. per lb. was given, but though abundant it was too troublesome a fishery to make it a pursuit, except when nothing else could be caught. †

It is very probable that the majority of our sharks have a very wide range, so that there is nothing of peculiar or local interest about them. The following list includes all that are known to occur in our seas:—

*Carcharias gangeticus*, or sea shark, and *Carcharias melanopterus* and *C. brachyurus;* the Whaler, *Galeocerdo rayneri;* Tiger Shark, *Galeus australis;* School Shark; *Zygæna malleus*, Hammer-headed Shark; *Mustelus antarticulus; Lamna glauca*, Blue Pointer; *Carcharodon rondeletii*, White Pointer; *Odontapsis americanus*, Grey Nurse; *Alopecius vulpes*, Thrasher; *Notidanus indicus, Scyllium maculatum, S. laticeps, Parascyllium variolatum, P. nuchale, P. ocellatum, Chiloscyllium furvum*, Dog-fish; *C. ocellatum, C. trispeculare, C.*

---

\* Fish oil absorbs oxygen so rapidly, and thickens to an extent which precludes its application to machinery.

† The flesh of sharks contains the most nutriment of any fish, but is hard, dry, and tasteless. From this reason it is perhaps less digestible. It is said that the Chinese use it as food, but this is true only of the fins.

PLATE XXVII.

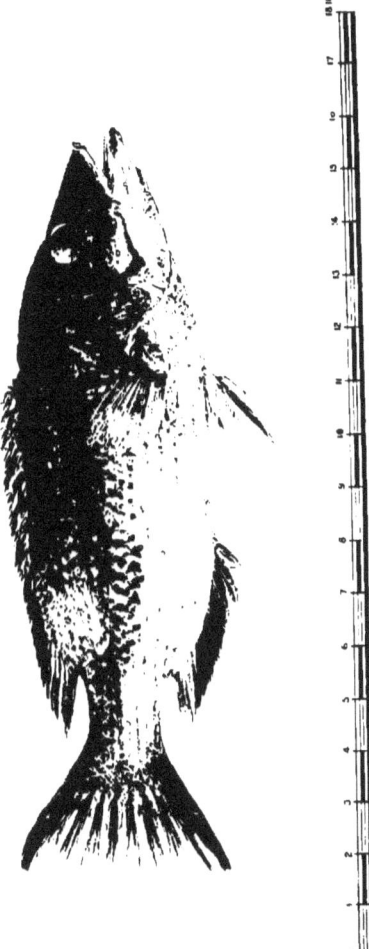

*Cossyphus uniuocellatus.*—GUNTH.

PIG-FISH.

*modestum;  Crossorhinus barbatus*, Wobbegong; *C. tentaculatus* (at Cape York), *Heterodontus Philipi*, Port Jackson *Cestracion* or shark; *H. galeatus; Acanthias blanvillei, A. megalops, Isistius braziliensis, Rhina squatina*, Angel-fish or shark; *Pristiophorus cirratus*, Saw-fish shark, *P. nudipinnis, Rhinobatus granulatus*, blind or sand shark.

In this list some of the species named have not been found at Port Jackson. They are representative species, which are, as far as known, confined to peculiar localities.

Some of Mr. Hill's remarks on these fishes are of such interest that an account of our fisheries would not be complete without at least a portion of them.

## The Sea Shark.

Of *Carcharias gangeticus* he says :—" This is about the largest of the sea sharks, and, like all sharks, ravenous to a degree at times. It arrives in this harbour about February, but at times later, in March, just according to the migrations of the sea-mullet, of which notice has been already taken. With these fishes it is found, and no doubt preys on them to an enormous degree.

"At this season—February, March, and April—it goes up the harbours and rivers, and may occasionally be caught with a very stout line and hook or harpoon. It is of great strength, and would pull a large boat about with ease. The sea shark gorges and gets rid of a great quantity of food, and one was caught in this harbour containing in its stomach (as far as my memory serves) the hind leg of a horse, from the hock downwards, with a shoe on the foot, eight or nine unsound mutton hams tied up in a gunny-bag, two ship's scrapers, and some sundries; notwithstanding all these he was ready to take more, and was captured.

"These cartilaginous fishes are very troublesome to fishermen outside the Heads, and frequently cause the men to leave the ground. It is astonishing to find how readily they will bite off the schnapper from the hook as you are hauling it in, leaving the head or perhaps the head and shoulders alone. Certain it is there are plenty of schnappers about them, and during the time you are fast to a shark the others would be diligent with their lines, as the fishes bite well at that time. I only wonder that sea-bathers are not more frequently bitten about February and March, as then they are very voracious and numerous. At the Figtree Baths a huge dog was purposely drowned. A sea shark coming by that way seized the dog by the middle, gave him a shake or two, and then swallowed him, and was afterwards swimming about with the appendage of a large stone hanging to his mouth. This did not appear to give him much trouble, for he was looking about for more food, and stood a few rattles on the head with an oar before he would retire to deeper water.

" The liver occupies a great space inside the shark, and is of value, for it yields a considerable quantity of oil which may be either rolled or boiled out; the oil from the former method is said to be of the better colour and gives less trouble. The other part of the shark is turned to no advantage generally, but the fins and tail, when cured and dried, are said to be of value among the Celestials, and are readily bought in the Chinese market; one of the uses they are put to there is for rich soups made richer by the addition of a fin of a shark boiled to a jelly. In this Colony they are sometimes saved, but the oil is the chief object when fishermen go out on purpose to kill sharks."

## The Wobbegong.

The same author gives the following account of *Crossorhinus barbatus* or Wobbegong:—"It is marked with irregular patches, and very like a carpet, having a brown ground. It grows to the extent of 4 to 5 feet in length. It has a large rounded head, and the mouth, which is of fair proportion, is situated near the extremity, with a slightly overhanging upper lip, armed with cirrus, which hang thickly around. The teeth are of two or three rows—small, sharp, and point inwards, so that the prey can scarcely escape when once fairly in its clutches.

"The wobbegong, of which there are several varieties on this coast, is chiefly nocturnal; but, like most other sharks, will not despise food in the day. At that time they are generally concealed under or at the base of overhanging rocks and patches of weeds, which it very much resembles. At night it travels about into clear sand patches, and does not move away from a sudden blaze of light—*that*, it is said, has rather the property of attraction to these sharks. It is lured readily by smell of fish, and it steals up with cat-like caution through the weeds, and poises itself on the nearest eminence, close to the spot whence the odour has arisen. This fish, except the gliding almost imperceptibly, has an awkward motion in munching the bait, or when it attempts to turn and move about. The stealthy manner in which it moves or lies in wait for its prey gives one the idea that it does not travel much in search but depends more on cunning.

"A party of us were spear-fishing one dark night at the Long Reef, by the aid of torches, accompanied by some aboriginals (who called this sport 'bidgee,' and further south 'chitmere'), when a wobbegong seized one of the blacks by the heel or lower part of his fustian trousers. To attempt to move it at that time is useless, and not till the fish makes a kind of gruff chuff, when he opens his mouth for the purpose, and offers a chance to get loose. However, the black, being only fast by the trousers, struggled to get loose, and did so by tearing a portion of them behind, which ran clean up into a slip to the waist.

"The aboriginals always look out for them among the reefs and dark corners when fishing with the torch. At Jervis Bay, one night, a black-fellow ventured out in the shallows for a considerable distance on this errand, and by accident placed his foot on or near the mouth of a wobbegong; the fish immediately seized hold, but the fellow never attempted to move but called out lustily for a light (his own torch had burned out), another black had started off armed with a good spear as well as with a light to his rescue; but during this time the fish had made the accustomed noise with a view to take a better hold, when the darkie availed himself to haul his foot out of its mouth, but in their alarm, and on coming ashore in haste, both he and the torchbearer fell down and they were both in darkness, but ultimately got safe on shore.

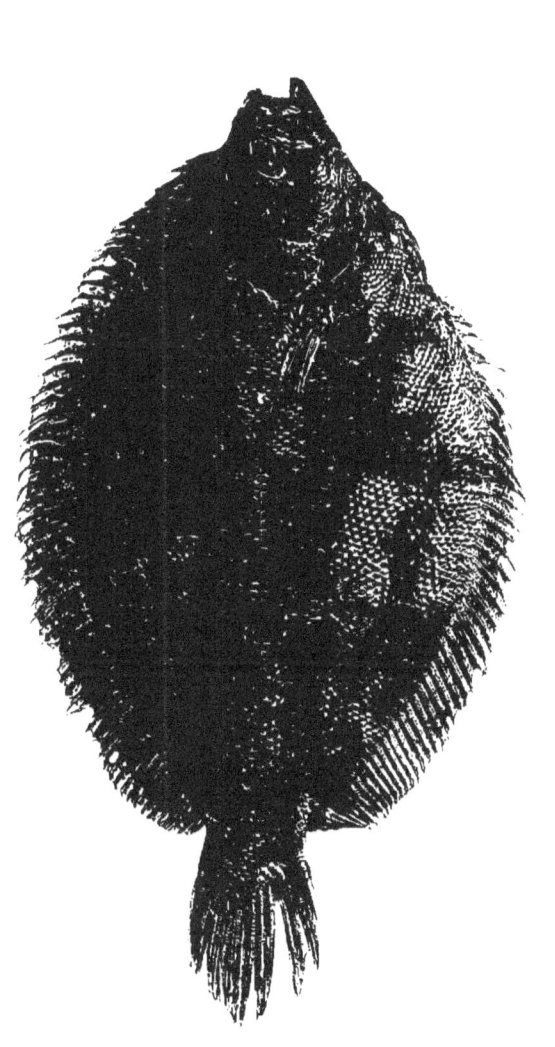

"The wobbegong is of little use; the liver will make some oil, and which is particularly good for parasites on animals, but the fish is not sought after, and the natives attack it when in their way, but do not meddle with it otherwise. South, as far as Jervis Bay, they call this fish thubbegong."

## The Blue Pointer.

*Lamna glauca* or the Blue Pointer is, Mr. Hill tells us, "not only a very powerful but perhaps the most bold of any of the large sharks visiting our shores. It grows to the length of 10 or 12 feet, but is not so big in the girth as the other. On the appearance of a 'blue pointer' among boats fishing for schnapper outside, the general cry is raised 'Look out for the blue pointer.' This fish exhibits cunning, and seizes the hooked fishes when they are near to the surface, at which time they are nearly exhausted, and bites them off just below the hook in an instant, leaving only the head, which has the appearance of being cleft off with a very sharp instrument.

"The great danger of these fishes is that in their speed and blinded by the eagerness of pursuit they may miss the object and go clean through the boat, or into it, as they have been known to jump high into the air. In such a case the whole crew would in all probability be destroyed by this one's attendants. Sometimes they have been known to leave their teeth in the hardwood of the keel or stern-post of the boat, no doubt having missed their aim, but when this occurs they make the boat shake from stem to stern.

"These are high-swimming fishes, and may be readily seen when about pushing their pursuits; the beautiful azure tint of their back and sides and independent manner they have of swimming rapidly and high among the boats in search of prey are means of easy recognition, and they often drive the fishermen away. They have a longer and sharper nose than the 'sea shark' or 'nurse,' and appear to be very active, and bold enough to come straight on after they have had a good blow; they appear to be very keen of scent, and altogether they are objects of interest. Various people in this as well as in other Colonies have been fearfully mutilated by sharks, sufficient in some instances to cause death very soon." Scarcely a year passes that some fatal occurrence is not recorded from sharks on some part of the coast.

## The Grey-Nurse.

The "nurse" here referred to is *Odontapsis americanus*, about which the same author gives many interesting facts. He says "it attains the length of 15 feet, with a much larger girth in proportion to other sharks; but the females which have been caught were not so long—8 to 12 feet—with an exception or two where they have been longer. There are incidents of interest connected with its season and capture worthy of note.

"This shark in appearance is comparatively short. That, however, is accounted for by its great girth, for I have known the length of a nurse to be 12 and its girth 7 feet. They have a longer and more pointed nose than the sea-shark, and the eye and nostril are in a straight line. The mouth, which is situated directly under the eye, is of a crescent shape, and the jaws are armed with four or five rows of teeth, formidable and sharp.

"These sharks are caught in October, November, and part of December, for their oil alone. After the latter month the liver yields little or none. During these months they yield a considerable quantity, the quality of which is good, and excellent for burning in lamps.

"These sharks used to be caught in Botany Bay, at regular 'nurse' grounds, near to the Seven-mile Beach, and what is known as 'Doll's Point,' where they could be seen of a calm day during the months of October and November, lying on the bottom in regular rows like logs of timber, and each row apparently as if they had been selected of one length. During these months, and on these grounds, they were caught with line and hook. Singular to remark that this operation was at one time systematically carried out, and the oil obtained from the liver of these sharks was used at the Tower (a Government station situated on the inner head of the north side of Botany Bay) for years, as well as by the fishermen of the village of that bay. The fact is also very curious that when fishing for these sharks you will, with a rare exception, catch them all of one length—no variation which can be detected by the eye; and out of twenty-eight caught in 1857, at one fishing, there was not the slightest perceptible difference. Occasionally a very large one was caught, and which forms the exception.

"In fishing for the grey-nurse, and when one of them takes the bait, the others rush towards it with a view either to participate in the food or to protect the one which has taken the hook, for they then overlie each other to that extent that it is like being fast to a rock; a constant and heavy strain alone will move the one you are fast to. When the battle commences, it gives some sport and plenty of work before you can get it alongside to enable you to give it a blow on the head with a club, which is part of the equipment for that purpose. After the oil season these sharks are very vicious, and move about freely; and when one is fishing for other fishes and happens to hook a 'nurse' on a good line, it will, with its great strength, drag the boat about, kellick and all, apparently without much exertion.

"The 'nurse' has been known to seize hold of the steer oar of a whaleboat when the boat has been moving rapidly through the water, and shake it with its teeth two or three times, let go its hold, and pursue and seize it again as if it was a living object, or, as it were, in derision or sport. At other times it has been known to attack the boat as if it smelled the bait within, or hover about, putting in an appearance every now and then. It was supposed that sharks could smell the living freight on board the ships connected with the negro slave ships in days of old; so with these, they have been known to seize hold of the sternpost and shake the boat till it quivered again."

*Heterodontus* or *Cestracion* has a peculiar interest attached to it, on account of its being of a form which prevailed in past geological times, or at least fishes very like it. The fossil forms exceed in size the species of our coast. They make their appearance in the Devonian period, increasing in the Carboniferous, and surviving up to the Chalk. The majority of the species however lived in the earlier secondary period.

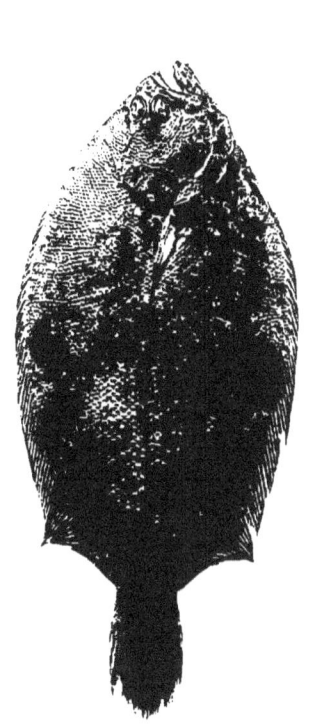

The Australian forms are small and harmless, as all its representatives were. Their teeth are only formed for crushing shells. Although, as Mr. Hill observes, "it is armed at the front of each dorsal fin with a strong pointed spine; the head is bony and hard, and the mouth is filled with a pavement of crushing teeth, curiously formed, and it seeks its food either amongst the crevices of rocks or on the ground, foraging about for shell-fish, cockles, crabs, and other *crustaceæ*, and at times will examine the edges of rocks just above water-mark in search of shell-fish, which its powerful jaws and teeth or plates are admirably adapted for crushing.

"It was supposed that Port Jackson alone had this shark, and it is the living representative of fish which once populated the Northern Seas, wherein the fossil remains are found in multitudes. It has since been found in many of the coast bays of Australia, and there are two distinct species—the one is much darker than the other in colour.

"I had occasion to catch it, and tried Long Bay and Botany Bay for that purpose, and was successful at both places. At Jervis Bay also I succeeded in shooting one in the head whilst in the position of searching the rocks with its head out of the water. They are also frequently hauled in with the net, and caught also with the hook and line, but are of no use, and only interesting from the fact that they are said to be the living representatives of what are found in fossil remains.

"This was one of the early sharks caught in this harbour, and figured in Governor Phillip's time."—E.S.H.

## The Shovel-nosed Shark.

*Rhinobatus granulatus* or shovel-nose which is properly speaking a Ray, is called here the blind or sand shark, though, as Mr. Hill remarks, it is not blind. He says "that it attains the length of 6 to 7 feet, and is also harmless, armed only with teeth resembling small white beads secured closely upon a card; it however can see tolerably well, and searches on sandy patches for *crustaceæ* and small shell-fish. It is much more active than the preceding shark, and can swim about rapidly, but it chiefly keeps on the bottom, and usually in company on particular grounds with *Cestracion*, and distinguished by the aboriginals from other sharks by the appellation of *erayoni*—without the teeth. It is occasionally brought in with the net of a night, but more frequently with hook and line."

## The Hammer-headed Shark.

None of the Hammer-headed Sharks of Port Jackson are of small size, but one specimen in the Brisbane Museum measures over 15 feet in length. It differs from other sharks in the shape of the head, which in some degree resembles a hammer. The orbits of the eyes project to a distance from each side of the skull, and are placed on either end of the hammer, giving to the fish a very odd appearance. It is armed with sharp teeth, and roves about at night near to the surface of the water, and when hooked it pulls hard and sheers about, no doubt to endeavour to cut the line.

## The Saw-shark.

The saw-shark must not be confounded with saw-fish, as their gill-openings are lateral not underneath. The snout is armed with an exceedingly long, flat lamina, with a series of strong teeth along each edge at right angles. From the position in which these teeth are placed it is problematical whether they obtain their prey by piercing or spitting with the point. "It is more likely that a sharp blow sideways would do greater damage in helping them to secure any living object. These saw-sharks were in great numbers on the coast about ten or twelve years back, and did great damage to the fishermen's lines. Although they might have been on this coast in their season for ages, still at that time they appear to have created some sensation, as they were much spoken of. In the north part of this island they are numerous, but I have not heard of any of large size."—E.S.H.

## The Angel-fish.

The Angel-fish is *Rhina squatina* of Linnæus, quite identical with the British species, though as Professor M'Coy remarks from its low power of swimming and habit of keeping on the bottom it is difficult to see how it could have spread over the area it occupies, or cross the immense depths of the ocean. It is very voracious, eating fish and mollusca, though never attaining anything above 5 feet in length. An excellent figure of this fish is given in Professor M'Coy's " Prodromus of the Zoology of Victoria," Part 4, Plate 34.

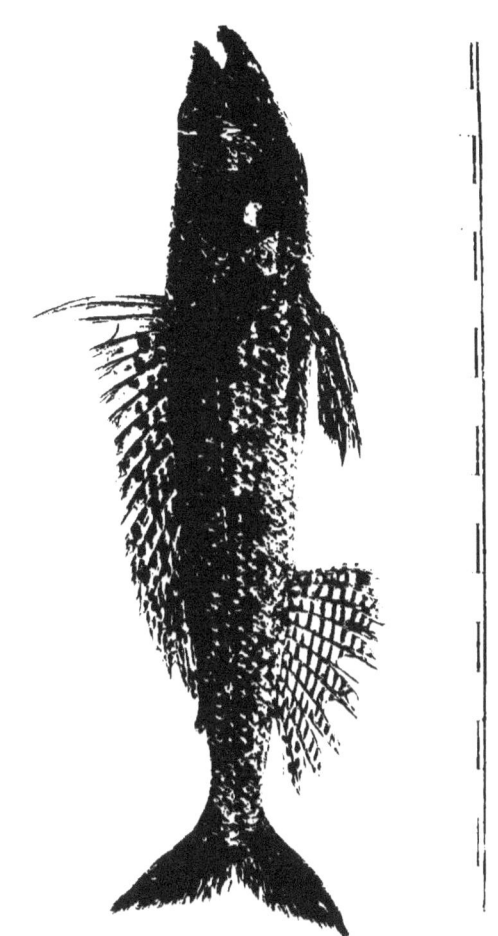

*Arripis purpurissatus.*—RICHARDSON.

THE SERGEANT BAKER.

# CHAPTER V.

## Rays.

IN the family *Batoidei* or Rays the body is excessively depressed, and forms with the expanded pectoral fins a somewhat circular disc, to which the tail appears as an appendage of varying length. In the families *Pristidæ* and *Rhinobatidæ* the habit is that of sharks, but the gills are placed underneath instead of at the sides. In all the anal fin is absent, and the dorsals are on the tail. True rays lead a sedentary life on the bottom; they swim by the fins, and the tail serves merely to steer them. The mouth is entirely at the lower surface, so that the prey is not seized with the mouth, but the fish darts over it, and holds it down with its body, and conveys it subsequently to its mouth.—G.S.F.

The Rays of Australia are:—*Pristis zysron*, saw-fish; *Rhinobatus granulatus*, shovel-nose; *R. banksii, R. thouini, R. dumerilii, Trygonorhina fasciata*, the Fiddler; *Narcine tasmaniensis, Hypnos subnigrum, Raia lamprieri*, Thorn-back, Sting-ray or Stingaree; *R. rostrata; Urogymnus asperrimus, Trygon narnak, Trygon pastinaca*, the Sting-ray of Sydney fishermen; *T. tuberculata, Urolophus cruciatus, U. testaceus, Myliobatus aquila, M. nienhofii, M. australis, Aëtobatis narinari, Ceratoptera alfredi*.

## The Devil Fish.

This peculiar and ugly name was applied to a stuffed specimen of enormous size by G. Krefft; it is in the Australian Museum. It has never been described, says Mr. Macleay, and now never can be, so much painting, puttying, and clipping have been practised in setting up the specimen. It was called the Devil Fish when it was caught in 1874, probably from some resemblance to the fish of that name in the Atlantic. The disc is very broad, in consequence of the great developement of the pectoral fins, which, however, leave the sides of the head free; and at the sides of the mouth are a pair of detached fins. This fish was caught in the vicinity of Watson's Bay, and deposited in the Sydney Museum. It affords a fine specimen of the family *Raia*, measuring across the wings or fins not less than 15 feet, and must have weighed something more than even the famous ones recorded by Cook.

## The Sting-ray.

*Tryon pastinaca* and *tuberculata*, our true Sting-ray, has a tapering tail, the first with and the second without a cutaneous fold, armed with a long arrow-shaped spine, serrated on each side. Pectorals united in front, body smooth or tubercular, teeth flattened. Mr. Hill speaks of one which he calls the Black Ray, though it is hard to say if it be the true Sting-ray or *Raia lampreiri*. The latter is not known to occur in New South Wales. He says that the Black Ray, "with a prominent head, is not only a powerful fish when hooked with a good tackle, but is of immense strength in its element; when on the bottom its form enables it to hold on, and it is like attempting to move a rock. On a line, when hooked, it exhibits great power, and often have I seen it towed into beaches to enable the amateur fisherman to get a firm and steady pull to land him. The angular form of its pectoral fins frequently reach the distance of 8 or 10 feet, and the weight exceeds 100, 200, or even 300 pounds.

"Rays in their season retire to deep water, and their habit is that of sharks, to which, as cartilaginous fishes, they are the gradation. At the breeding season they approach the coast, and enter our bays and harbours. In some of the species the large and well-developed pectoral fins form a broad flat disc, with a thin and slender tail."

The weight of some of these fishes may be readily understood by quoting from Captain Cook, when he visited Botany Bay in 1770:—

"'He had observed that the large sting-rays, of which there is a great plenty in the bay, followed the flowing tide into very shallow water. He therefore took the opportunity of flood, and struck several in not more than 2 or 3 feet of water. One of them weighed no less than 240 pounds after his entrails were taken out. The next morning, as the wind still continued northerly, I sent out the yawl again, and the people struck one still larger, for when his entrails were taken out he weighed 336 pounds.'"

With regard to the spine or sting with which these fish are armed, though it inflicts a ragged and even a most dangerous wound, yet whether it is a poisonous one or not has not been clearly established. Dr. Günther unhesitatingly says that it is, and this is the almost universal belief. I have seen two or three cases of such wounds, and the agony complained of was too great to attribute to the puncture alone. Yet no serious consequences followed once the pain had subsided.

## The Torpedo.

Our Torpedo or Electric Ray is *Hypnos subnigrum*, that of Tasmania is *Narcine tasmaniensis*. The difference is in the teeth, which are flat in *Narcine* and tricuspid in *Hypnos*, while the latter has the nostrils round and open, and in the former they are confluent with a quadrangular valve. I quote from Mr. Hill a few facts concerning this remarkable fish, which has an electrical apparatus at each side of the head. It is a small fish, rarely exceeding 40 lbs., which for a ray is trifling.

In Middle Harbour, on the clear sand bottoms, at night, they are frequently hauled in with the seine. "On one occasion, when we were more inclined to play than to assist with the net, we amused ourselves with a large torpedo which had been hauled in. Four of us, after feeling a shock individually, joined in a circle holding hands, and those at the extremity touching the fish; simultaneously we completed the circle and experienced a shock similar to that from a galvanic battery. On another occasion, and at another place, we endeavoured to continue some experiments on one of these fishes which had been caught, but it was so enfeebled that the shock was scarcely perceptible. This we attributed to having expended its power upon the net with which it had been in constant contact during the capture. In about half an hour afterwards the fish had apparently somewhat recovered, as the shocks were appreciably stronger, but ceased with death."

PLATE XXXVI.

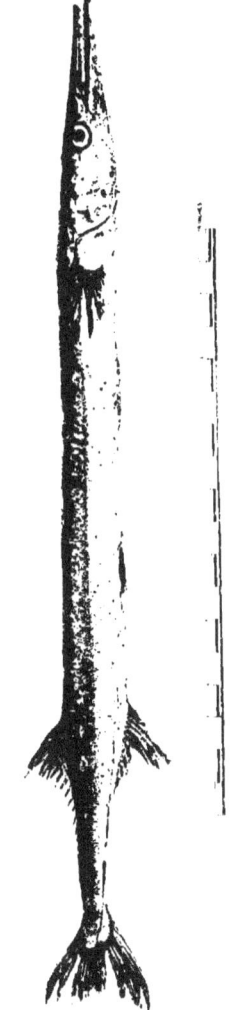

*Belone ferox.*—GÜNTH.
LONG TOM.

The organ by which this shock is given is thus described by Cuvier:— "The space between the pectoral fins, the head, and the gills, is occupied on each side by an extraordinary apparatus, formed of little membranous tubes placed close to each other like a honeycomb, and subdivided horizontally into minute cells, abundantly supplied with nerves. It is in this apparatus that the electrical or galvanic power resides which has rendered these fishes so famous, and from which their name is derived; with it they can inflict violent shocks, and they most probably use it to deprive their prey of power."

## The Fiddler.

The Fiddler, or *Trigonorhina fasciata*, otherwise a harmless fish, is a great pest to the fishermen at times in the harbours, readily and quietly taking the bait intended for some other fish of a more useful character. Like all the shark tribe, they have a very powerful scent, and seize the bait by overlying it, with a mouth which is well underneath. These fishes, if not very cautious in their movements at this time, are struck by the line, and the hook very often pierces some other place than the mouth on the under part of the body.

All the stingless Rays or Raiidæ are useful as food, but they are never eaten in this Colony.

# CHAPTER VI.
## Our Fresh-water Food Fishes.

In such a dry climate as that of Australia, where the rivers are few, and except in winter mere sluggish narrow streams, it is not to be expected that our fresh-water fishes can be either numerous or abundant. For the colonists in the immediate neighbourhood they have a certain economical importance, but the fisheries do not employ many hands, nor are they ever likely to. The river system of the western side of the Dividing Range is confined to the Murray and its tributaries, of which only the Darling, Murrumbidgee, Lachlan, and Macquarie, can be called rivers, and of these some of the upper feeders run dry in seasons of severe drought. Outside the influence of these rivers, and in the far western country, such a thing as a fish is never seen. In the little creeks small cat-fish (*Copidoglanis tandanus*) is found. This affords the only sport that anglers can obtain, and the fish, one of the *Siluridæ*, forms a delicacy which is appreciated by those to whom fish is such a rarity. With regard to the fish fauna of the waters of the western side of the range, it is very uniform; the same genera and species are found in all of them. If there are any local peculiarities they are as yet unknown. Foremost amongst them as an article of diet is the "Murray Cod," of which there are two species; they belong to the family of Perches and to the genus *Oligorus*. It is distinguished by having an oblong body covered with scales. The cleft of the mouth is rather oblique, the lower jaw being the longer. Teeth viliform, extending to the vomer and palatine bones, but with no canines. There is one long dorsal fin, the first eleven rays of which are spinous. The anal fin has three spines, and the tail is rounded. The preoperculum has a single margin, which is smooth or faintly toothed. Some of these fishes grow to an immense size, and they are found in the sea as well as rivers. One, *O. terræ-reginæ*, Ramsay, goes by the name of the Groper in Queensland. In the Brisbane River specimens are caught weighing over 160 lbs., and measuring more than 6 feet in length. A large species named *O. gigas* (the Hapuku of the Maoris), is caught off New Zealand, which reaches a weight of 100 lbs.

## The Murray Cod.
(Plate XLI.)

*Oligorus macquariensis*, Cuv. and Val., which is the "Kookoobul" of the Murrumbidgee natives, "Pundy" on the Lower Murray. In this species the height of the body is four times and three quarters in the total length, the length of the head three and a-half, the diameter of the eye is one seventh of the latter. Preoperculum, supra-scapular and preorbital entire, pectoral and ventral fins short; fifth dorsal spine the longest; second and third spine of the anal fin nearly equal; colour greenish brown, with numerous small dark green spots; belly whitish, but the colour varies much in different places.

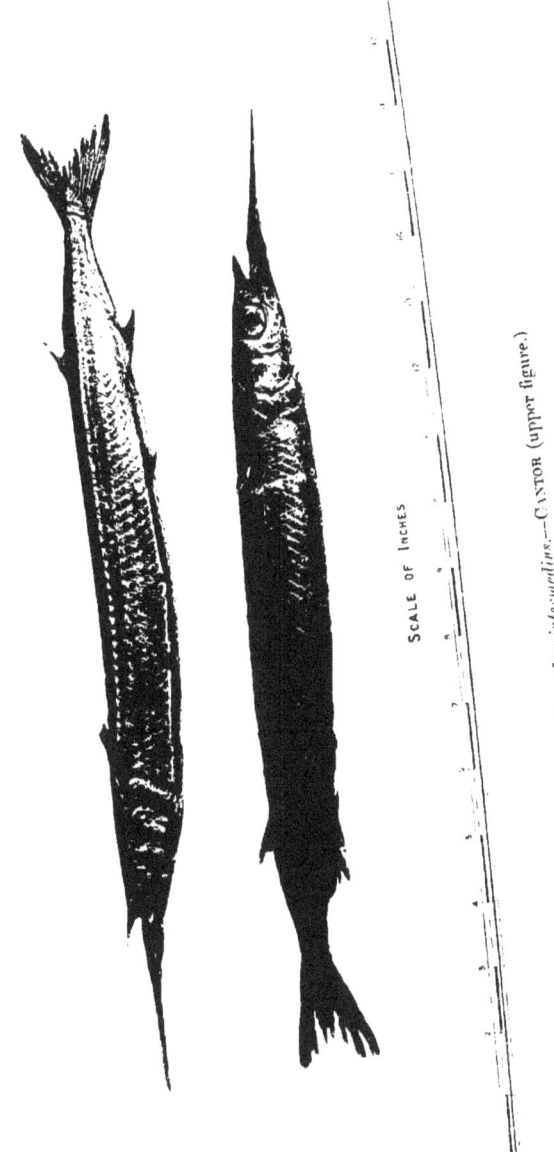

*Hemirhamphus intermedius.*—CASTOR (upper figure.)
COMMON GARFISH.

*Hemirhamphus regularis.*—GÜNTH (lower figure).
RIVER GARFISH.

*O. mitchelli*, Castelnau, differs from the above in having the head much broader and the eye considerably larger, it being one-sixth the length of the head, the upper jaw longer than the lower, the operculum more rounded, and the caudal fin a little longer. The height of the body is also three times and a third in the length without the caudal, instead of four times and a-half, as in the other species, and the upper profile is more convex. It occurs in all western rivers, but *O. macquariensis* is found in a few of the eastern rivers to the north of Sydney. The colour of *O. mitchelli* is a livid grey, entirely covered with small round obscure spots. The fishermen call this the Murray Perch, to distinguish it from the Cod. The smaller specimens are more spotted.

Both fishes are of the same voracious character, devouring every fish or animal of any kind which their enormous mouths can compass. They are both excellent fishes for the table, and have been known to reach a weight of 120 lbs. The young fish are found at the very head of the streams, the old ones generally lower down in the deep reaches of the rivers; and it is said that, like all other fishes, they have periods of migration, appearing to ascend the streams in summer and to descend them in winter. Very little art is required in the capture of these fish—a strong line, a large hook, and a lump of beef for bait are all the requirements necessary. The spawning season is in midsummer, Mr. F. A. Tompson, of Wagga Wagga, making the time about November, and Mr. Warren, a fisherman at the same place, positively affirming it to be January, and in this view Mr. Leitch and others coincide. Mr. F. A. Tompson has seen the fish, as he believed, actually depositing the spawn—one fish, the female doubtless, moving along the bottom of the water, forming a furrow in the sand with its chin, while another fish (the male) closely followed in its wake. Mr. Warren never saw the fishes spawning, but had often found what he believed to be the spawn attached to logs, and he affirms that he can discriminate perfectly between the spawn of the cod and the other percoid fishes of the Murrumbidgee.—R.R.C.

## The Golden Perch.

The Golden Perch or Yellow-belly represents two species of *Ctenolates*, *C. ambiguus*, and *C. christyi*, a species described by Count Castelnau, from the Edwards River. The first of these is common in all the rivers and lagoons of the interior. It is a very rich and delicate fish, and attains a weight of 7 lbs. or more; its time and manner of spawning is the same as the cod. The spawn is believed to be hatched in a fortnight after deposition. When fresh this fish is coloured very beautifully. The body is of a magnificent green, the sides and behind the dorsal, the upper parts of the body, are rich golden. The head is a beautiful mixture of green, purple, yellow, and scarlet, with fine golden tinges; the belly is white, the dorsal fin purplish green, anal scarlet, with its base yellow and its end purple, pectorals scarlet at their base, and yellow in their second half; the eye is purple, with an interior white ring. These colours are subject to great variation, and the belly is sometimes red. The young fish have little of the fine hues of the adult, and they are much more elongate. The head is purple, and the dorsal fin is grey, bordered with black.

In *C. christyi* the upper parts are brownish purple, and the lower white. It is very rare. It must be mentioned that Günther and Castelnau place these fishes in the genus *Dules*. The main distinction from *Oligorus* appears to be that the scales are small and strongly ctenoid, the spines in the dorsal are ten in number, and the preorbital and preoperculum are finely serrated with small denticulations on the lower limb of the latter. *Dules auratus* is a common synonym. *Ctenolatus* was a genus erected by Günther for this species, which he then called *C. macquariensis*, but it would seem now as if the genus was abandoned by him, while Mr. Macleay thinks it should be preserved.

## The Silver Perch.

The "silver perch" or "bream" (*Therapon richardsonii*) is the perfection of fishes, extremely rich and delicate in flavour. It frequents running streams more than the last-mentioned fish, which is often found in lagoons and billabungs, and it affords good sport to the angler. A full-grown fish attains a weight of 5 or 6 lbs. It is not caught often with the hook, the very small size of its mouth preventing its taking the hooks in common use.—R.R.C.

The genus *Therapon* is very extensively represented in Australia. It is partly marine, and partly fresh water, and spread over the whole area of the tropical Indo-Pacific. *T. therops*, *T. servus*, and *T. cuvieri* are very common in Australia, and extend from the east coast of Africa to Polynesia. They are readily recognized by the blackish longitudinal bands with which they are marked. All the spines are small. The genus has an oblong compressed body, with scales of moderate size. Viliform teeth, those of vomer and palatine rudimentary or absent. Dorsal fin with depression in upper margin, twelve or thirteen spines, the anal fin with three. Preoperculum serrated. The average length of *T. richardsonii* is about 11 inches. The preorbital is very strongly serrated, the preoperculum is rounded, armed with a series of long spines posteriorily, and shorter ones below. The operculum has two spines, the lowest being the longest; spines of the anal very long, particularly the longest; colour, greyish blue, lower parts dirty white, sides yellow, scales bordered with black, head bluish, lips and lower parts rosy, eye yellow, first dorsal dark, rays purple, the second dull yellow below, black above, caudal black, anal with purple rays, ventrals white, rays rosy, pectorals yellow at base, black above. Found in all the rivers of the Murray system, and called Kooberry by the natives. To distinguish it from the other species, the catalogue of Macleay must be consulted, as they are very numerous.

There are several other percoid fishes of good size and quality found in the Murray River system, such as *Therapon niger*, *Murrayia güntheri*, *cyprinoides* and *bramoides*, and *Riverina fluviatilis*, all described by the late Count Castelnau. There are besides several species of Percidæ of small size, which are useless except as food for other fishes.

*Murrayia* is a genus placed by Count Castelnau between *Dules* and *Therapon*. It has eleven dorsal spines, the operculum is denticulated in all its length. A line of small teeth on the palatine bones, the caudal fin is rounded, scales minutely serrated, and the head cavernous. It is peculiar to Australian rivers.

*Riverina* resembles *Murrayia*, but has twelve spines in the dorsal fin, and is without teeth on the palatine bones. It is found on the Murray, but rarely.

The professional fishermen at Wagga Wagga, on the Murrumbidgee, and at Albury and Echuca, on the Murray (the first supplying Sydney with fish, the others Melbourne), use, we are told, a net with a long pocket in the middle, which

PLATE XXXVIII.

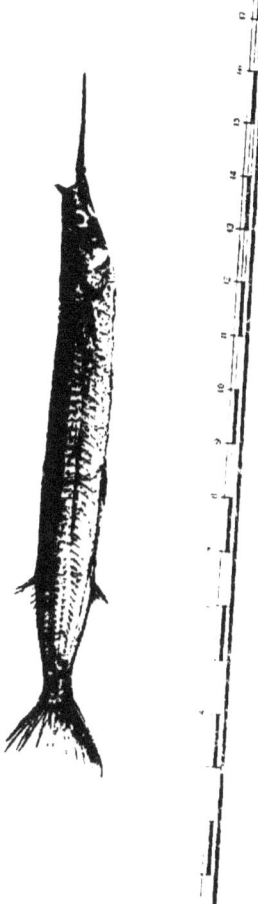

*Hemiramphus regularis.*—GÜNTH.
RIVER GARFISH.

they stretch across the river at night, thereby intercepting fishes passing down the river. If, as mentioned above, these river fishes have their migratory periods, the complete stoppage of the river by nets may be most injurious or even destructive to them.—R.R.C.

## The Freshwater Black-fish.

A very common fish in some of our rivers, both of eastern and western waters, is the "Black-fish," *Gadopsis marmoratus*, which belongs to a genus of the most extraordinary fishes known. It is, according to Professor M'Coy, an intermediate type between the *Acanthopterygious* fishes, in which the anterior rays of the dorsal fin are simple spines, with the scales usually ctenoid, and the *Malacopterygious* fishes, in which all the rays are soft and branched, and the scales usually cycloid. In *Gadopsis* the cycloid scales, the general form, the imperfect filamentous jugular ventral fins, and the majority of the characters so nearly agree with *Malacopterygions* that all the most recent writers with Dr. Günther class it with the Anacanthini, although the anterior rays of the dorsal and anal fins are distinctly spinous. Our species is a mud fish, and attains the length of 16½ inches, but it is generally caught by emptying the water-holes where the summer heat has made them low. It is good eating, but like all these mud-fishes, very rich and oily. In *G. marmoratus*, Richardson, the head is one-fourth of the length. Prof. M'Coy has described two species. One, *G. gracilis*, in which the head is proportionately much shorter. This is found in the river Yarra, and differs in its habits from the fish of western waters, as it is readily caught with a line. It is also a far better fish for the table, and is much esteemed in Melbourne.

There is a second species of *Gadopsis* (*G. gibbosus*, M'Coy), which is proportionately shorter, deeper, and with a much more convex dorsal outline. It abounds in the Bunyip River, Gippsland, Victoria. Another distinction is that it has twelve instead of ten spines in the dorsal. The colouring of these fishes is very variable. Some are light olive green, becoming yellowish-white towards the belly, the sides, back, and fins being mottled irregularly with dark brown. This marbling varies from brown to olive in smaller or larger patches, and the yellow varies to orange. It is only when the mottling is very dark and thick that the name Black-fish is at all applicable. Sometimes the brown marbling is slight and distinct, and the general colour yellowish olive. The scales vary also from truly cycloid to an indented margin, and undulating lines of growth approaching the ctenoid type. An excellent figure of *G. gracilis* is given in M'Coy's "Prodromus of the Zoology of Victoria," decade 3, plate 27, fig. 2. The name *Gadopsis* is meant to mean like the cod or *Gadus*.

## The Freshwater Cat-fish.

The "cat-fish" (*Copidoglanus tandanus*) is very abundant in the lagoons and back waters of the western rivers, and is said to be a most excellent fish, but there is a very general prejudice among Europeans against its use: it is very fat and eel-like in flavour, and averages, when full grown, 2 feet in length. In this species, as in most if not all of the *Siluridæ*, the ova are fertilized by the male fish before leaving the body of the female, and both sexes seem to unite in the subsequent attendance on the nest in which the ova are deposited.—R.R.C.

o

The genus belongs to the *Siluridæ*, which is distinguished by a short dorsal fin in front, with a pungent spine. A second long dorsal is united with the caudal and anal into one fin; teeth in the upper jaw conical, on the vomer molar-like, in the lower jaw mixed; barbels eight, cleft of mouth transverse; eyes of moderate size, with a free orbital margin. The gill-membranes are separated by a deep notch, united anteriorily only, the united portion not attached to the isthmus. The first branchial arch with a fringed membrane along the hinder edge of its concave side. Ventral fins, many rayed; head rather compressed. There are two other genera in Australia, *Plotosus* and *Neoplotosus*. The differences are principally in the gill-membranes and teeth. They are found in the fresh waters of Western and Southern Australia. The catfish proper (*C. tandanus*) is brought abundantly to Melbourne by the Echuca railway. The fishermen distinguish three sorts, the white, black, and blue, but they are only varieties. The colours are subject to the greatest changes. The back is steel colour or olive, often marbled. The head and barbels pink. Again, there are specimens of a brownish-black, and others white, with the head yellow. The teeth are white, in a triangular band on each maxillary, twice as broad as long, and the vomerine in a semi-circular disc. The first dorsal arises from the neck, and is nearly as high as the body, and its spine is more than half the length of the head; pectoral as strong, but shorter.

## The Bony Bream.

A fish of the herring tribe is also found in these rivers (*Chatoessus richardsonii*). The native name on the Murrumbidgee is "Ka-i-ra," and by the white settlers it is sometimes known as the "bony bream." It is said to appear at times in immense shoals. It is a handsome, good-sized fish, but almost useless for food on account of the multitude of bones in it.

In *Chatoessus*, which belongs to the herring family, the body is compressed and the abdomen serrated; scales of moderate size; snout obtuse, or obtusely conical, more or less projecting beyond the cleft of the mouth which is narrow and without teeth. There are ten species known from the brackish and fresh waters of Central America. One species ranges to New York; the rest belong to Australia, the East Indies, and Japan. We have two species, *C. richardsoni* and *C. erebi*. The first is a more convex and less elongate form and has the last dorsal ray half the length of that of *C. erebi*. Count Castelnau states that it is much esteemed as food in the Melbourne market and sells at a high price; also, that according to Blandowski it is called *manur* by the natives. It leaps frequently out of the water and is easily caught by its elongated ray in thin fine nets, laid by the natives horizontally on the water. It is most numerous on the Darling. In June and July it is at its best, and then forms one of the principal articles of food for the natives. The young women are not permitted to eat them, from a belief that if they did all the fishes would die. It is a very fat and nourishing fish. It is also placed on the tops of graves for superstitious purposes. Mr. Macleay does not think these statements correct. The name with the Wooradgererie or Murrumbidgee language

*Anguilla australis.*—Richardson.
AUSTRALIAN EEL.

was Ka-ec-va. The fish was not common, and considered too bony to be good as food. The flavour is good enough, but the bones make it useless. The average size is 10 to 14 inches.

## W. Australian Herring.

*C. erebi* has been found (says Count Castelnau) in Western and North-western Australia, as well as in the Norman and Brisbane Rivers, Queensland. In the latter place it goes by the name of the sardine. Mr Bostock, of Swan River, says that it is known there as the "herring," quantities being smoked with *Banksia* wood or sawdust and sold. It is said to be excellent eating, but not likely to become of much economical importance, as its habits are not gregarious.

The fishes of the eastern rivers are of a less exclusively fresh-water character than those of the western. We have the "perch" (*Lates colonorum*), frequently found in salt water, the "mullet" (*Mugil dobula*), visiting the sea annually after attaining the adult state, and the same has been found to be the case with the "eel" (*Anguilla australis*). The "herring" of these rivers also (*Clupea novæ-hollandiæ*) is frequently found down in salt water. These four species are in all the rivers of the east coast, and are all good fish—the perch has been mentioned before among the salt-water fishes; and also the mullet, the eel, and the herring. In addition to these there are in the Cox, Nepean, and other tributaries of the Hawkesbury, a species of *Apogon*, several species of *Eleotris*, and a *Centropogon*. Further north, in the Clarence and Richmond Rivers, we find an additional mullet (*Mugil petardi*), an additional perch (*Lates curtus*), and what is very remarkable, two of the fishes of the Murray system—the "cod" (*Oligorus macquariensis*), which is got in some branches of the Clarence, and the "cat-fish" (*Copidoglanis tandanus*), abundant in the Richmond.

## The Apogon.

*Apogon* is a small fish of the perch tribe, and belongs to a genus of brilliantly coloured "coral fishes." Very few species out of nearly 100 are found out of the tropics, and fewer still are fresh-water. Our species (*A. güntheri*) is about 4 inches long, of brownish pink colour without spots or bands. On each side there is also a golden tinge, and all the scales are covered with minute black dots. Mr. Macleay thinks it is *A. novæ-hollandiæ* Val. It occurs in Port Phillip and Tasmania, as well as Port Jackson.

## The River Gobies.

*Eleotris* is one of the Gobies, from which the genus differs only in not having the ventral fins coalescent. They are, for Gobies, rather large fishes, tropical, and more fresh-water than marine. Some occur in the inland waters of the African continent. Our species is *E. australis*, Krefft, a fish of about 5 inches in length, occurring in all the eastern rivers and creeks. It is of a yellowish-brown colour, covered with minute black spots in five or six longitudinal lines. The tail is spotted, and the base of the pectoral fins a bright yellow; there are also some faint bands on the second dorsal. The head is scaly to the obtuse snout, the lower jaw prominent, and the teeth in viliform bands.

*Centropogon robustus*, Günth., is in all our eastern rivers. It belongs to the *Scorpænidæ* family, and a genus which is distinguished by having no groove on the occiput, no pectoral appendages and no cleft behind the fourth gill. The species is a small brownish fish, marbled with black. Its scales are small, and the fourth and fifth dorsals are the longest. None of these three fishes are of any commercial importance whatever. It has already been shown what peculiarities this fish has under the name of the "Bull-rout."

## GALAXIADÆ.

In the upper and shallower parts of the creeks and rivers rising in the Blue Mountains one or two species of *Galaxias* are found. They are cylindrical fishes of 8 or 10 inches in length, and without scales, inhabiting only the colder rivers of the Southern Pacific. They are probably good for the table, but they are rare and for the most part small. They are known to some as the "Mountain trout." In describing a new species from Mt. Kosciusko, *G. findlayi*, the Hon. W. Macleay makes the following remarks (Proceed. Linn. Soc. N.S.W., vol. 7, p. 106) :—"I received from Baron von Mueller, a few days ago, two specimens of a small fish which inhabits the icy pools of the snowy range in the neighbourhood of Mt. Kosciusko. The Baron writes as follows :—' I saw the same little creature in several of the waters high up in the Alps, during my exploration of the Snowy Mountains in 1853-4, and 1855, and again in later years when travelling, but I was in the then pathless alpine regions, unable to preserve zoological specimens. When in 1874, I for the second time ascended Mt. Kosciusko, I saw this species of fish again in the little glacier ponds, but missed catching any, my time being so much occupied, during my brief stay on the snowy summit, in the pursuit of plants.' The two specimens now to be described were captured by S. Findlay on Mt. Kosciusko. They are both small, the largest not exceeding 3 inches, and evidently immature." Macleay then refers to a former paper read before the same Society (vol. 5, p. 45), describing another species of that genus, from the head waters of the Colo River at Mount Wilson, and he pointed out the probability of fishes of this kind being abundant, and of considerable size in the cold snowy streams of the Australian Alps. He remarked in the same paper that though such fishes were found in the upper tributaries of the Grose, at heights of two or three thousand feet, none were found in the Nepean and Hawkesbury, into which these streams flowed. The author attributed this not so much to the falls as to the difference of temperature, and mentioned that its distribution showed it to be essentially a cold-water fish. The family is a remarkable one, containing only two genera, *Galaxias* and *Neochanna*. Both are small fresh-water fishes, only found in the Southern Hemisphere. *Neochanna* is a remarkable mudfish of New Zealand, which is caught in burrows which it excavates in clay or consolidated mud, at a distance from the water. It is, says Dr. Günther, a degraded form of *Galaxias*, from which it only differs by the absence of ventral fins. These fish have no scales or barbels, they have a thick lip, and the ova fall into the cavity of the abdomen before exclusion. Altogether the family is a most isolated one, having no relationship with any other. "The species

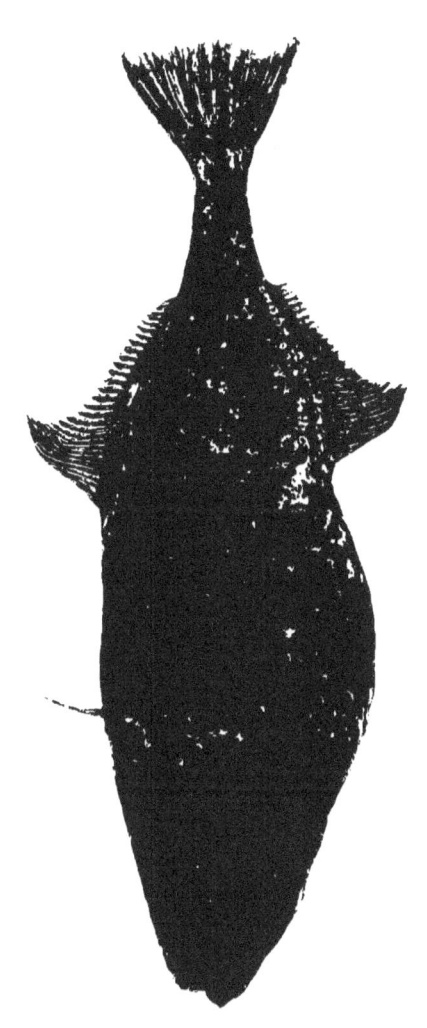

of *Galaxias*," says Mr. Macleay, "are numerous, but so much alike, that it is, looking at their distribution, more than probable that they are one and all only permanent local varieties of the same fish.

"But the chief interest attached to these fishes is in their distribution. They are found only in the rivers of Southern Chili, Magellan Straits, the Falkland Islands, Tasmania, New Zealand, and those parts of Australia where the rivers take their rise in the Snowy Mountains or in cold elevated table-lands; so that in fact we find this singular fish in all the lands which extend into the colder regions of the Southern Pacific and nowhere else. The deduction from this singular fact is very plain. At one period,—probably very remote even in a geological sense,—the area of land above the sea in the antarctic regions must have been very much in excess of what it is at present, at all events sufficiently extended to admit of some kind of continuity across the whole width of the Pacific between the southern extremity of South America and Australia. There is no other way of accounting for the appearance of these fishes in such widely different localities."

The subject thus suggested is a very large one, and is worthy of attention in connection with the whole of the Australian fauna. The agency of birds in transporting ova must not, moreover, be lost sight of, though there are difficulties in the way of that explanation as well. It should be mentioned, moreover, that some of our Australian freshwater mollusca have a very wide range in the Pacific, while the marine molluscan fauna is comparatively restricted. The spread of ova by means of birds is the usual way of accounting for this.

## The Australian Grayling.

Though not a fish of New South Wales, it may be as well to mention here the Australian Grayling, which in character, habits, and the manner of its capture is almost identical with the English fish of that name. In shape there is some difference between the two fish. The local Grayling is smaller, of a more uniform thickness, and with a less prominent back-fin; but it possesses strongly the peculiar Grayling odour. A newly-caught fish smells exactly like a dish of fresh-sliced cucumber. It is widely distributed in Victoria, and very abundant in all the freshwater streams of Tasmania. It is the only native fish which affords any sport to the fly-fisher; but it prefers a red worm, and can be best taken with the coarsest tackle. It seldom exceeds a pound in weight. As it has a well marked adipose fin it was referred to the *Salmonidæ*, but it is now placed in the family *Haplochitonidæ*, a group of fresh-water fishes which represents the Salmonoids in the Southern Hemisphere. Only two genera are known, viz. :—1. *Haplochiton*, fresh-water fishes, abundant in the lakes and streams near the Straits of Magellan, coasts of Chili and the Falkland Isles. It has the general habit of Trout, but without scales. 2. *Prototroctes*, with the habit of *Corregonus*, scaly and with minute teeth. The Australian Grayling, which used to be known as *Thymallus australis*, belongs to this genus, and is called *Prototroctes marœna*. In Melbourne it goes by the name of the Yarra Herring. There is another species in New Zealand.

## CHAPTER VII.

### Oyster Fisheries.

THE oyster fisheries of New South Wales are of immense value, and may become more so when we remember that "natives" are now ten guineas a bushel in London. It may soon be worth our while to export them, especially as ice would not be wanted. London pays between four and five million annually for oysters, and double the quantity might be sold. In Paris oysters are 12 francs the 100. The "tinning of oysters might easily be conducted at a profit amongst us."

According to Dr. J. C. Cox (Pres. Lin. Soc., N.S.W.), who has given special attention to the subject, we have five distinct species of oyster in New South Wales; these are—*O. angasi*, Sow., mud oyster, *O. subtrigona*; Sow., drift oyster, *O. glomerata*, Gould, rock oyster, *O. circumsuta*, Gould, a rare species, *O. virescens*, Angas, also a rare species. The two last are of no commercial value, being rare, of small size, and difficult to remove from the rocks.

### The Mud Oyster.

*O. angasi* was, and is still by many regarded as only a variety of *O. edulis*, Linn., of European seas. The differences are that the valves are dentate at the margins. Mr. Sowerby says that the sculpture is less coarse, and Mr. Angas says that it is larger but more regular. Such distinctions will, however, hardly bear specific classification. It is found rarely in Port Jackson, though once very common, and it has grown very scarce in Tasmania. In Port Lincoln (Spencer's Gulf, South Australia) it is still abundant. It was formerly abundant in New South Wales, as the kitchen-middens of the aborigines show; it was also formerly abundant in Pliocene times, as there are raised beaches in Victoria and South Australia entirely composed of the shells of this mollusc. It is undoubtedly one of the best if not the best of oysters. The species or variety *O. rutupina*, Jeffreys, is abundant in Tasmania. This is the "native" or Colchester, or Carlingford oyster of Britain.

### The Drift Oyster.

*O. subtrigona.*—Shell subtrigonal, oblong, or subquadrate, ponderous, rather narrow towards the umbones, broad at the ventral margin, quadrate, margin strongly plicate, lower valve deep, greenish white, edged slightly with purple, radiately plicate outside, concentrically banded with fawn and purple, hinge acuminated, sides crenulate near the hinge. The sculpture of the shell is bold and large, and the square character of the ventral margin is striking.

This is the Drift Oyster, so called because it is believed that its beds are shifted by the influence of storms or tides. It lies in beds consisting for the most part of free unattached individuals, or attached to masses of drift matter, or to each other by adhesions of the lower valve. The

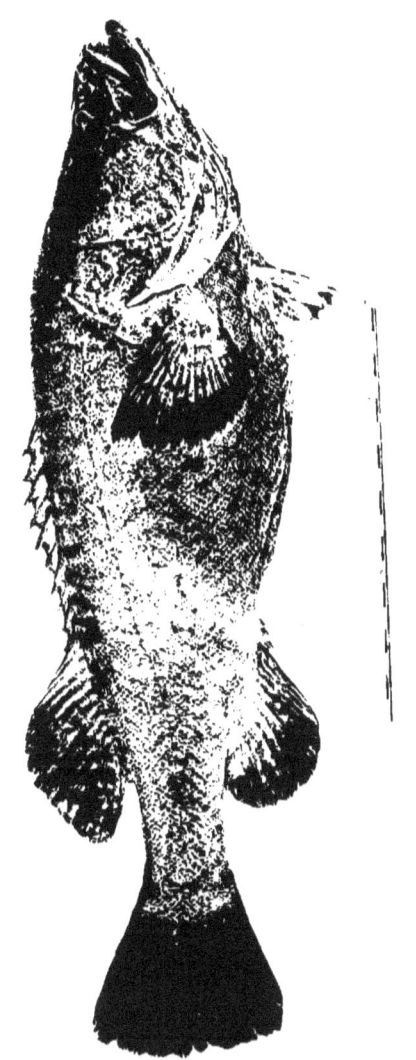

*Oligorus macquariensis.*—Cuv. & Val.

MURRAY COD.

beds are always in moderately deep water, and never uncovered by the tide. It is the common oyster of New South Wales and Brisbane, and is largely exported to the neighbouring Colonies.

It is commonly supposed, says Dr. Cox, that this species and the rock oyster are identical, and with this view many persons have gathered rock oysters and planted them in beds with a view to their cultivation. He maintains, however, that this only leads to loss, for if this species is placed in positions where it is uncovered by every tide, it speedily dies; but if properly cultivated on the beds it usually inhabits, it will become a most important article of food.

Mr. Oliver and many others maintain that the drift oyster will flourish well if left under the same conditions as the rock oyster, that is where it is uncovered by every tide. The first-named gentleman tells me that he has tried the experiment, and succeeded in rearing a fine lot of drift oysters. On the other hand, rearing rock oysters in drift oyster beds is an experiment successfully and frequently effected by every oyster grower. Altogether the evidence preponderates in favour of the theory that *O. subtrigona* and *O. glomerata* (that is the so-called rock and drift oysters) are the same species under different conditions. With ordinary care and attention they may be grown in deep or tidal waters.

## The Rock Oyster.

*O. glomerata*, or Rock Oyster. This is the kind which is found so abundantly adhering to the rocks on all the east coast of Australia. Dr. Cox regards it as distinct from that which occurs so abundantly on the Queensland coast, especially on the coral reefs, where all the outstanding blocks of coral (nigger heads) are covered with them. This is *Ostrea mordax*, Gould. Having given the subject every consideration in travelling along the east coast I cannot regard the species as distinct. No definite specific distinction can be found when large numbers of species are examined. This oyster is generally found adhering to rocks which are not always covered by the tide. It is often beautifully frilled, and of a rich purple colour. "If," says Dr. Cox, "individuals of this species are placed in proper trenches or in positions where clear fresh sea-water will flow over them at each tide, they thrive and fatten to an extent which makes them a valuable article of food and an important commercial product, and by placing low stakes of wood or other material for the spat to adhere to when emitted from the mother shell, they are easily and successfully propagated. But when placed in such positions, especially on mud flats which are uncovered by every tide, they are liable to the attacks of a number of other molluscs, and unless the water which flows over them is pure, and free from decomposed vegetable matter and from grit, they suffer from the irritation caused by such particles, some discolour and waste and others die." To this species is also to be referred *O. cacullata* of Born. The variety is merely an overgrown form of the hinge very common on the coral reefs.

A Royal Commission was appointed in the latter end of the year 1876 to take the important question of the Oyster Fisheries into consideration, and a very excellent Report from that Commission was brought up and printed early in 1877. The Report was accompanied by the draft of a Bill, which it was hoped would have been taken up by the Government and passed into law, but unfortunately no further steps were taken in the matter.

There is no part of the world better, we might almost say so well, adapted for the growth and culture of the oyster as New South Wales. The climate, the nature of the coast line, with its numerous inlets and creeks, and the natural existence of the best varieties of the bivalve itself, all combine to mark this country out as the most likely of all places for successful ostreiculture. Unlike the coasts of England, Scotland, and Europe generally, where the coldness of the temperature in summer is sufficient to prevent sometimes the shedding of the spat at all, and always to limit the quantity of it, here we have invariably the spat distributed with a profusion which under proper regulations would make our oyster supply unlimited and inexhaustible. And yet with all the advantage of a bounteous supply we are actually informed that the natural beds are so nearly exhausted that a bag of oysters can now only with difficulty be raised in a day, where a year or two ago it would have been easy to dredge 50 bushels in the same time. The same process of exhaustion, with a few praiseworthy exceptions, is going on in the leased beds, and we can see no remedy for this crying evil except by making it more to the interest of those leasing oyster beds to conserve, improve, and keep up the supply, than to clean the beds out in the shortest possible time and have done with them.—R.R.C.

Many of the suggestions made above have been already carried into effect by the Fisheries Act of 1880, as will be notified in its proper place.

A few words on the possibility of oyster culture in the manner which is so successful in pisciculture may be inserted here. The list above given of the species or varieties of oysters in Australia is very nearly all that we know on the subject, and that is very little. Of that little, much is uncertain. Many experienced persons do not regard the rock oyster as distinct from the drift oyster, as I have already mentioned; they say that the difference arises merely from the manner in which it is reared, and to this opinion I am disposed to subscribe, for reasons which have appeared in the previous remarks. One would think that there is not much to be said about the organs or parts of an animal so simple in structure as the oyster, but this is not the case. Perhaps it is for want of a knowledge that can be easily obtained that so little has been done, and this I propose to remedy in a few explanations which are necessary in order to understand what I have to say on the matter.

Oysters are spoken of as being fat and lean, and their shells are described as upper and lower, all of which terms are a source of confusion. There is no fat in an oyster. What is usually known as such is the ovary or milt, that is to say, the roe or unfertilized eggs of the male, and the soft roe or fertilizing fluid of the male. These organs form the great mass of oysters which are in good condition. Oysters are of distinct sexes. It has been suggested that there may be an altenation of the sexual character, and those which are males one season may be females another. Such things are not unheard of in the animal kingdom, and there is some evidence in its favour in the case of oysters. On the whole, however, the weight of evidence is against it. These remarks apply to the habits of the European and American species. Of the Australian species we know nothing, except from analogy. They are nearly allied to oysters in other parts of the world, and may even be specifically identical. What is therefore true of one may be true of all. Before proceeding further, definitions of terms must be given.

The oyster has, properly speaking, no head or tail. The portion in which the mouth is situate is called the head. It is nearer to the hinge than the outer edge of the shell, and therefore the hinge side is called the anterior and the free edge of the shell is the posterior side of the body. In the young state the two shells of the oyster are of equal size and alike, but it soon becomes attached by one side, and then the adhering side grows much faster and is much larger. This lower valve is invariably the left one, and it is deep and concave, while the right valve is usually flat. If an oyster be held edgewise, so that the right valve is on the right, and the hinge away from the observer, the dorsal surface will be uppermost, and the ventral below. The hinge is provided with an elastic ligament which would always keep the shells a little open were it not for the strong adductor muscle in the middle of the inside of each shell. This has to be cut through before an oyster can be opened. Both valves are lined with a fine transparent membrane called the mantle. Inside this there is a fringe with a narrow dark margin; this fringe will be found to consist of four laminæ, which occupy the ventral half of the valves, and extend from the open edges nearly round to the hinge. Remembering the definition of what is the dorsal and what the ventral edge of the body, it will be found that the gill laminæ are free on the ventral side—that is distinct from each other, but they are joined together and united with the body and mantle on the dorsal side. Round the adductor muscle the great mass of the substance is taken up by the ova or milt. Just by the side of the muscle is a small transparent vesicle like a blister; it will be seen moving from time to time in a freshly opened oyster, but very slowly and at long intervals. This vesicle contains the heart and the movement is its pulsation. The blood of oysters is colourless. The transparent veins and arteries can be easily seen on a close inspection of the tissue. The heart consists of a small compact white ventricle and a loose spongy transparent auricle. The auricle receives the blood from the gills, and the ventricle pumps it to the various organs. The whole process can be easily seen with a very ordinary hand-lens. In front of the gills—that is between them and the hinge—there are four flaps of flesh, two on each side of the body. These are around the mouth, which is near the hinge, and away from the open end of the shell. The oyster feeds on extremely minute organisms, as the mouth is supplied by the currents caused by cilia. These are organs only visible with high microscopic powers, and constantly moving. They are like little hairs. The intestine is coiled up amidst the ovary, and there is a long slit close to the middle of the posterior face of the adductor muscle, which is the anus.

As already observed, the sexes are divided in the oyster, but there is no external mark by which a male oyster can be distinguished from a female. The only test is the microscope. If a portion of the white stuff called the fat be taken out and mingled with a drop of sea water, and then covered by thin glass and examined with a good inch objective, it is easy to say at once whether the oyster is a male or female. If a male, the white fluid is clouded with minute masses of granules which can scarcely be distinguished. If a female, the ova are quite distinct as regular granulations. It requires a very high magnifying power to make anything out of the male fluid, but with a good quarter of an inch objective it is seen to be filled with small organisms like tadpoles, with tails which are incessantly moving.

P

By mixing the male and female fluid it has been found perfectly easy to fertilize the ova. After a period varying from one to two days the eggs are hatched; a minute free-swimming organism is the result. By the aid of warmth and many favourable circumstances eggs may be hatched in two hours. A somewhat warm temperature in the sea is the best for hatching, and the slightest cold of a severe character destroys the ova. Our climate should be about the best for oyster culture, and as far as the rock oyster is concerned the more abundant it becomes as it is traced northwards and into warmer seas. About twelve hours after the free-swimming stage the shells begin to appear, with the germs of other organs. The shells are distinct from the first, and for a long time have a smooth rounded outline. At the end of six days the shells nearly cover the embryo, which is hardly more than visible to the unassisted eye as the merest grain. All the larger organs can be seen with a microscope, as they are then separated, with the exception of course of the reproductive tissues such as ovary and milt. After swimming about in this state for a few days, the period of which has not been exactly ascertained, the young oyster, the so-called "spat," fastens itself by one valve of its shell to any rough and clean body. It is then very thin and delicate, but grows fast. If it has not some object to fasten to it sinks to the bottom. There it may find the living and dead shells of its relatives or a rock or other shells, and there it will grow. It is not uncommon to find nearly a hundred small "spat" on one oyster shell. It is then that oyster cultivators collect them and lay them in convenient places to fatten. Without artificial assistance there can be no doubt that the great mass of oyster ova perish.

Many observers have asserted that the ova are fertilized within the ovary and are afterwards nursed in the folds of the mantle. Dr. Brooks has proved that this is not true of the American oyster (*O. virginiana*), and it is probably untrue of all. I have carefully examined many ripe female oysters, and never could find any embryos on the mantle. It is probable that the egg is discharged into the water, to be fertilized, where in oyster-beds and amidst its own species it easily may meet with a male cell. If however it does not meet with such, the ova are rapidly destroyed by sea-water. The fertilized ova crowd together on the surface as a thin scum, where they no doubt fall a prey to many enemies, small ones of course, because they are too minute themselves to form an article of food to any but very little creatures. Still the chances of their escape from accident are not in their favour. This is why preserves and ponds must be so advantageous to their cultivation. If the collection of the spat is found to be so profitable, how much more would properly secluded places be where the ova could be left in tranquillity with all the chances of life in their favour. Artificial fertilization is not wanted,—all that is required is the protection of hatched spawn. Möbius* estimated the average number of eggs laid by *O. edulis* at over a million (1,012,925). Eyton rates this higher, and says that there are about 1,800,000.† Dr Brooks calculates from a much more accurate series of observations, and estimates the ova from the largest Virginian oyster as about 60 million, or for an average one about ten

---

* "Die Austern und die Austernwirtschaft," von Karl Möbius, Berlin, 1877.

† "History of the Oyster and Oyster Fisheries," by T. C. Eyton, London, 1858. The figures above are given at p. 24.

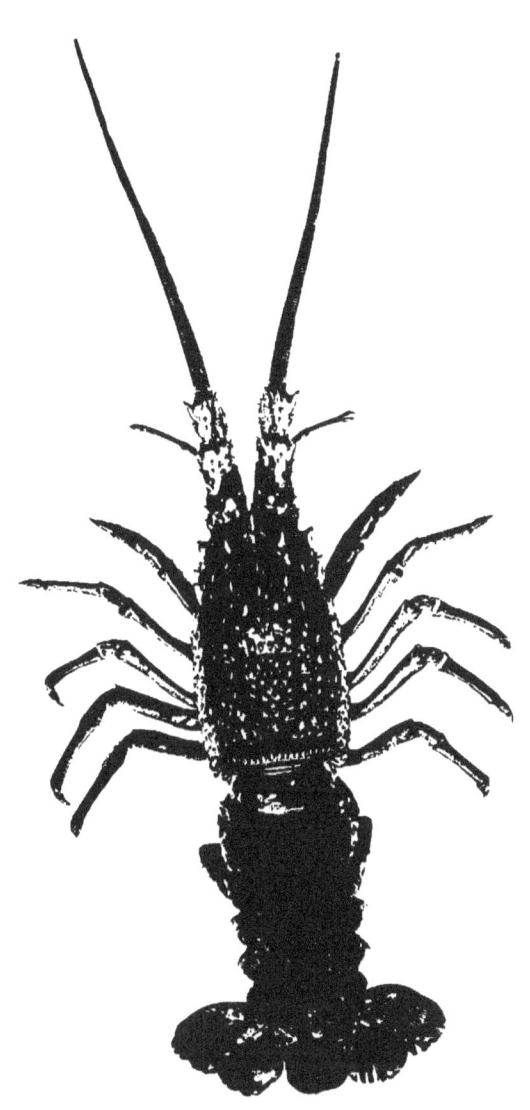

million.* So that, as far as ova are concerned, there is no want of material, and the male sperma are of course still more numerous: all that is wanted is protection. With regard to the methods by which this can be obtained, I condense the following from Bertram's "Harvest of the Sea" (p. 237), who states that the secret of there being only a holding-on place required for the spat of the oyster to insure an immensely increased supply was found by the French. Probably they were indebted in some degree to the methods pursued in England (Colne and Whitstable) for the idea. The plan was simple enough. Strong pillars of wood were driven into the mud and sand, arms were added, the whole was interlaced with branches of trees, and various boughs were hung over beds on ropes and chains, whilst others were sunk by weights in the water. A few boat-loads of oysters being laid down, the spat easily found a home. In 1850 the oyster beds of France were nearly exhausted. St. Brieuc, Rochelle, Marennes, and Rochefort could no longer supply the market. Oyster culture was begun at the Island of Ré where there are few trees, but stones and tiles were largely used as collectors for the oyster spat. The work was begun in 1858, and ten years afterwards there were upwards of 4,000 "parcs" and "claires" as they are called, constructed on the foreshores of the island. The system was inaugurated by a stonemason, named Bœuf. He enclosed a portion of the foreshore of the island, about 30 yards square, with a wall of rough stones about 18 inches high, and in this enclosure or parc he laid down a few bushels of growing oysters. In the course of a year he was able to dispose of about £6 worth of oysters. Elated with this success he doubled the size of his enclosure, and more than doubled his profits, without in any way encroaching on his breeding stock. The news of this success induced the French Government to cede portions of the shore to be used as oyster parcs by labourers, who in return paid a nominal rental of a franc a week. The first duty that had to be performed was to clear off the mud, which is fatal to young oysters, and which abounded in that locality. After this, rocks had to be blasted to obtain stones for the parc walls; then these had to be built, and the parcs stocked with breeding oysters. In a short time an incredible number of oysters was gathered on shores which a few years ago were of no value.† "So that this branch of industry now realizes an extraordinary revenue, and spreads comfort amongst families which were formerly in a state of comparative indigence. A series of enormous and unproductive mud banks, occupying a stretch of shore about 4 leagues in length, are now so transformed and the whole place so changed as to appear to be the work of a miracle.

Dr. Kemmerer, of St. Martin's, Island of Ré, has invented a tile which he covers with some kind of composition that can, when occasion requires, be peeled off, and this plan is useful for the transference of the oyster from the collecting parc to the fattening claire.

* "Development of the American Oyster," by Dr. W. K. Brooks, p. 15. Printed as an Appendix to the Report of the Commissioners of Fisheries of Maryland, January, 1880, Annapolis, State Printers, 1880.

† The remainder of this extract is from the Report of the Oyster Culture Commission, Sydney, 1877, p. 10, &c. It is abridged from Bertram's "Harvest of the Sea," to which I have added a few facts here and there of later date, taken principally from Simmond's "Commercial Products of the Sea."

Lake Fusaro is highly interesting as being the first seat of oyster culture. It is the Avernus of Virgil. It is still devoted to the highly profitable art of oyster-farming. The mode of oyster-breeding at this place is now, as it was eighteen centuries ago, to erect artificial pyramids of stones in the water, surrounded by stakes of wood, in order to intercept the spawn. Fagots of branches were also used to collect the spawn, which must find a holding-on place within forty-eight hours after its emission, or it will be lost for ever.

The Royal Commission (Ireland) say :—Hurdles and fascines have been found to answer well as collectors, and they will be found cheaper. They are fixed in rows, by means of pegs, about 2 or 3 feet above the oysters, which are scattered on the soil under them.

Furze bushes are also found to answer fairly, but fascines and bushes are scarcely so suitable in a tide-way, in consequence of the liability of the twigs to catch weed, break, and float away, when the spat is carried with them. In all cases when wood is employed for collectors, it should be dry, hard, and sapless, and cut, at least, in the preceding season. Oysters are more easily detached from wood collectors; the loss or damage to the shell breaking them off is least upon fascines, as the twigs are easily broken off; the loss is greater on hurdles, greater still on tiles, and greatest of all on stones. The young oyster, though somewhat malformed at times on twigs, soon regains its shape when detached without damage. Tiles are largely used in France because they are cheap—about £2 per thousand. One cultivator, at Auray, possesses 200,000 tiles, and on these he obtained, in 1869, six millions of oysters.

In New South Wales the production of oysters is immensely beyond our present requirements, and nature has also provided us so amply with holding-on places (rocks, mangrove trees, &c.) for collecting the spat, that it appears almost superfluous for us to allude to the subject of oyster-breeding; but this state of things may not always continue, and at some future time information on breeding oysters will be as useful as that on the growth and fattening of oysters is at the present time.

As respects the fattening of oysters.—The nature of the bed or soil on which it rests is a matter of the greatest importance. Bertram says the beds of 'natives' are all situated on the London clay or on similar formations. * * * The portion of the beds set apart for the rearing of 'natives' is as sacred as the waxen cells devoted to the growth of queen bees. But, although called 'natives,' in many instances they are not 'natives' at all, but are, on the contrary, a grand mixture of all kinds of oysters, being brought from Prestonpans and Newhaven, in the Firth of Forth, and from many other places, to augment the stock. Many circumstances highly favourable to the growth and fattening of oysters are the reverse for successful breeding. Growth and fattening will proceed where there may be a large amount of fresh water and a strong current: the former would prove prejudicial to spatting, and the latter tend to prevent the adhesion of spat—at least in the locality at which it is voided. It is a remarkable fact that there are no fine-flavoured oysters where there is not fresh water, and this fact was noticed by Pliny more than eighteen hundred years ago. The Royal Commission (Ireland) says: For fattening there are few places better than a salt

PLATE XLIV.

*Astacopsis serratus.*—SHAW.
FRESHWATER CRAY-FISH.
1.—Animal slightly reduced in size. 1a.—Head. 1b.—Claw. 1c.—Tail. All natural size.

marsh. The fattening ponds (termed claires) at Marennes and La Tremblade, of which sketches are appended, are at both places formed out of salt marshes, and are in many instances only old disused salterns or salt-pans in which rough salt was made. The number of oysters laid down in claires is proportioned to the time it is intended they should remain there; for as the food of the oyster is limited, a smaller number will of course fatten more rapidly than a larger number. The average distribution is about two or three to the square foot. The oysters thus fattened are of excellent flavour and quality.

Mr. Cholmondeley Pennell, Inspector of the English Oyster Fisheries, who was sent by the Board of Trade to inspect and report upon the French modes of oyster culture, says in his report :—' The fattening pits (claires) are excavated from 1 to 2 feet deep, and are of all shapes and sizes, from 10 to 60 yards square, which latter is the maximum, the usual size being from 40 to 50 yards square. It is in these pits that the celebrated green oysters are fattened. Round the margin of the claires, at Marennes, a trench or channel is excavated a yard or two wide, and an extra foot deep, the object of which is to equalize the temperature when the shallower water becomes too hot or too cold. One portion of the side of each claire is cut down to the depth at which it is wished to keep the water; this depression communicates with the nearest gully or natural channel, and at spring tides (when only the water in the tides can be changed) the tide, winding its way up the channel, finds ingress and egress. The same channel also serves to carry away the waste water whenever it is wished to lay the pits dry, for which purpose the simple method is adopted of digging a hole in the clay bank, which is readily stopped up again when desired.'

During the summer months the sea has free ingress and egress to the claires to purify them, and the coating of blackish mud which has collected on the surface during the preceding year is also removed. In August they usually stop up the gaps in the banks, in order that the continued action on the soil and water may produce the greenish creamy scum with which the surface mud of all the claires is covered. Oysters in the claires do not begin to fatten until late in the autumn and winter. A large quantity of oysters will live well in the pits, but they will not fatten if too numerous. There is no doubt that the fewer oysters that are placed in a pit, the more food there will be for each of them and the quicker they will fatten. Wherever these claires have been constructed they have succeeded, and, when once constructed, the labour and expense of working them are small. The claires at Marennes occupy a strip of low-lying clay country on the river Seudre. The soil is marl, that is, a mixture of chalk and clay, and is of various colours—greyish, blue or black, greyish yellow, and in some cases red. The muddy or marly bottoms are most favourable to the growth and fattening of the oyster. Professor Sullivan says :—' The soil of all places successful as oyster-fattening stations contains more or less of a fine flocculent highly hydrated silty clay, abounding in vegetable and animal matter, derived chiefly from diatomacea, rhizopoda, and other microscopical organisms; and that the soils of those places which have proved successful as breeding-stations always contains some of it, but not necessarily as much of those which fatten; and lastly, that in those places which have proved failures, this peculiar kind of mud is either wholly absent or inferior in quality and quantity.'

The Royal Commission (Ireland) say, fruitful oyster mud may vary within very wide limits, from almost pure sand to almost plastic clay. In the very sandy grounds, there must however be always a sufficient quantity of highly hydrated clay to render the sand adhesive and to preserve it from becoming a mere loose running mass.

In the clayey grounds there must always be calcareous mud to make the clay porous and prevent it becoming too hard,—clay marls, with some intermixed sand, being perhaps the best of all materials for oyster grounds.

The earth known as the London clay appears to be the soil peculiarly adapted for oysters. It may be well here to explain that the term 'London clay' is employed in a general and a special sense. In the former it is used as a collective name for a number of beds of the old tertiary formation, consisting of gravels and sands below and of clays above. In the special or limited sense, it is applied to the bluish or blackish clay, sometimes mixed with a greenish coloured earth and white sand, which forms the upper parts of the beds just mentioned. London clay is plastic clay, not differing much in chemical composition from ordinary potter's clay. All fruitful oyster muds contain organic matter, always due in part to the presence of infusoria, and sometimes in part to small algæ or confervæ, remains of shell-fish and other marine creatures.

Bertram says, one of the most lucrative branches of oyster farming in France is the fattening of oysters in claires, at Marennes, which have been brought from the Ile de Ré breeding parcs. In the claires the oysters become green, and of considerably more value than the white oyster. The peculiar colour and taste of the green oyster are imparted to it by the vegetable substances which grow in the claires. The industry carried on at Marennes consists chiefly of the fattening in claires; and the oysters operated upon were at one period of their lives as white as those which are grown at any other place; indeed, it is only after they have been steeped a year or two in the muddy ponds (claires) of the river Seudre that they attain their much-prized green hue. The ponds (claires) for the manufacture of these green oysters—the oyster *par excellence*, according to all epicurean authority—require to be watertight, for they are not submerged by the sea, except during very high tides. Each claire is about 100 ft. square; the walls for retaining the waters require, therefore, to be very strong. They are composed of low banks of earth, 5 or 6 feet thick at the base, and about 3 feet in height. These walls are also useful in forming a promenade, on which the watchers or workers can walk to and fro and view the different ponds. The floodgates for the admission of the tide require also to be thoroughly watertight and to fit with great precision, as the stock of oysters must always be covered with water, but a too frequent flow of the tide over the ponds is not desirable, hence the walls, which serve the double purpose of both keeping in and keeping out the water. A trench or ditch is cut in the inside of each pond, for the better collection of the green slime left at each flow of the tide, and many tidal inundations are necessary before the claire is fit for its stock. The oysters placed in them are a year or sixteen months old, and it is two years at least before they are properly greened.

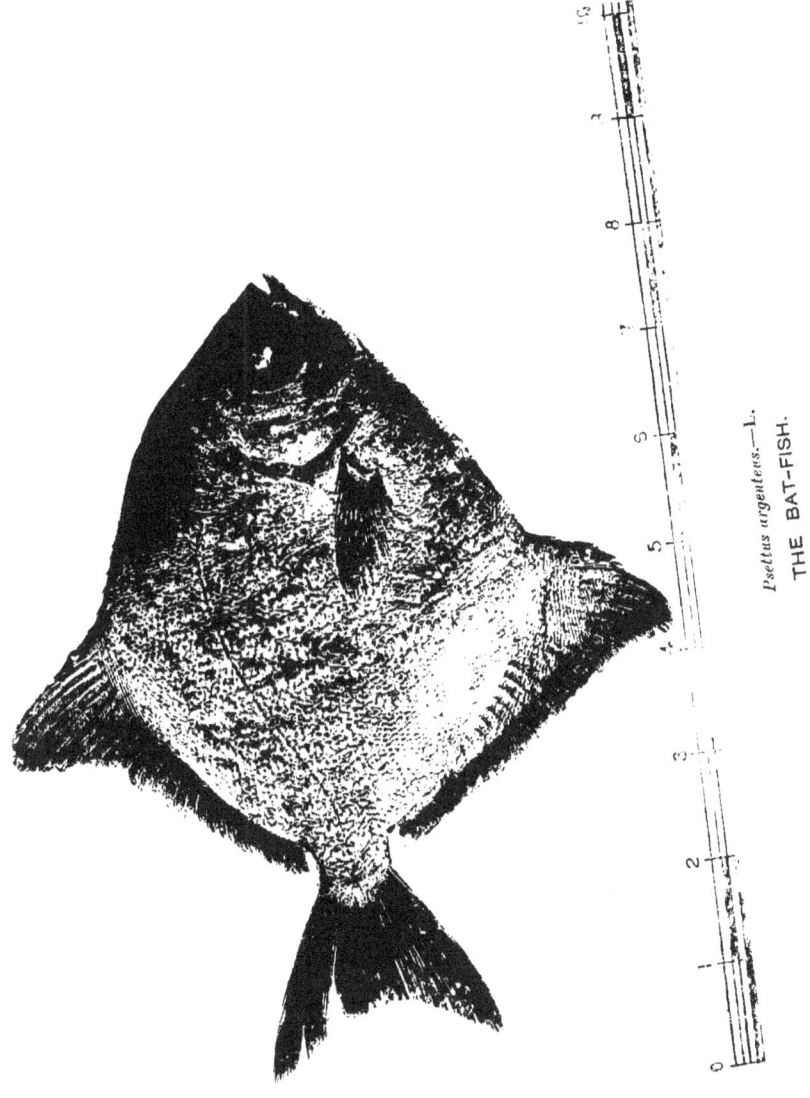

*Psettus argenteus.*—L.
THE BAT-FISH.

Auray in Brittany is, next to Arcachon, the seat of the most important of all French oyster fisheries. There is one establishment in the Auray district which comprises about 100 acres in a single enclosure and about 12 hectares between the enclosure and the sea. In 1864 the sea broke in and submerged it, causing as it was thought great destruction; but the proprietor took advantage of the accident to form the parc into an oyster tank by means of substantial and costly embankments. In 1876 the owner laid down six million oysters, more than half of which were about $1\frac{1}{2}$ inch long. All have grown well, and so satisfactory have been the results that contracts were entered into for the supply of a million of oysters to Paris, and the same to London, and the quantity is now probably doubled to each place.

In this Colony we have plenty of salt-water swamps, marshes, mud-banks ("crossets") more or less covered by the tides, where the fattening of oysters, or "greening" them, could be carried on. We have the same kind of rich mud as the estuary of the Thames, which is so celebrated for fattening oysters. Analyses made of the mud from one of the bays of George's River showed it to be similar to the London clay out of which Portland cement is made. Our unsightly and unhealthy waste marshes might then be made a great source of wealth. The "claires" need not be of any particular shape or size. Oysters grow best in shallow water. A fattening ground is usually a small creek with muddy banks, and the bed is made in the middle with shells on which the oysters are laid. Mr. Holt (Chairman of the Oyster Commission) constructed many claires on the coasts of New South Wales, more than 30 miles in total length, but his experience was decidedly against damming them. At first he made many flood-gates and dams, according to the most approved systems he had seen in France; but he has since done away with them entirely, and let the oysters have the full benefit of the ebb and flow of the tide. This has been a great saving in the expense, and the oysters have done better. It should, however, be mentioned the ebb and flow is much smaller here than on the coasts of France, and with consequently less danger of the stock being carried away.

Oysters have many enemies, one of the commonest of which is the sponge (*Hymeniacidon celata**), which forms for itself branching cavities by its siliceous spiculæ, and completely honeycombs the shell. It is thus easily crushed, and the animal becomes exposed and dies. It is said that exposing the shell to the sunshine is a remedy to this, and in effect the rock oyster, which is often long uncovered by the tide, does not seem to be much attacked by this parasite.

"Ten or twelve years ago the oyster industry in France was in a high state of prosperity, but five or six years later it was in a most deplorable state, and Mr. Pennell and others gave melancholy accounts of the failures of oyster culture in that country. This did not arise from natural causes, such as frost, snow, floods, &c., which occasioned such

---

* This was the old opinion, and is quoted by Buckland. Dr. Bowerbank contends that the mischief is done by an annelid worm which leaves borings which the sponge afterwards inhabits. *H. celata*, Bk. (which Buckland refers to as *Cliona*) is in its anatomical details one of the simplest and smallest of sponges, being a mere thread.

tremendous losses of spat and oysters in England, Ireland, and Scotland, but from the negligence and greed of the cultivators. The Royal Commission (Ireland, 1870) say in their Report: There is no reason to doubt that the decline in production (in France) is to be attributed to the neglected state of the collectors, and also to the selling of too many of the parent oysters, and thus annihilating to a considerable extent the source of spat. This, the Commission say, is admitted by the proprietors themselves, who have found their expectations to get spat without parent oysters to be delusive, and they are now taking means to renew the stock of oysters and collectors. The selling of their breeding oysters is but a repetition of the old story of killing the goose that lays the golden eggs. But now the tide has turned, and the French having learned by bitter experience not to trust solely to their fine climate and great natural advantages, have put their shoulders to the wheel, and by skill and industry have turned the bountiful gifts of Providence to good account. Like causes can never operate in New South Wales to injure the oyster industry, from the fact that there are many localities where marketable oysters cannot be profitably dredged for consumption, but where spat can at all times be obtained in any quantity.

"Mr. Farrar, Secretary to the Board of Trade, in the evidence he gave before the said Select Committee of the House of Commons (1876), said :—'Mr. Pennell was sent by the Board of Trade, in 1868, to inspect the French oyster fisheries, and he gave a most melancholy account of them—nothing could be worse. The Irish Commission confirmed that melancholy account; but now it appears from the official returns of the French Government (1876) that the production has enormously increased. At Marennes the private cultivators have been enormously successful; at Cancale the value of the oysters produced had risen from 97,375 francs in 1869 to 720,800 francs in 1874. The octriculturists, who have established parcs on the banks of the Auray, gather considerable quantities of young oysters in their collectors, and many of them have already realized important profits. Many of the proprietors of parcs are embarrassed by the abundance of their produce.' Mr. J. A. Blake, Inspector of Irish Fisheries, in the evidence he gave before the said Select Committee (1876) said :—'French oysters will cause a great revolution in the oyster trade in England; so that we need to care very little about our own production at all, but look more to the fattening.'"

The official value of the oyster produce in France for 1873 was given as follows :—2.477,565 francs. The dredging in the ports of Granville, Cancale, and L'Orient, produced in 1873 nearly 13 million oysters, against over 4 million and a half in the previous year; in 1874 it was 13 million and a half. From the official statement lately published, the following statistics are given for the commerce in oysters in France, for the season from the 1st Sept. to 30th April :—

| | Oysters taken from the beds. | Value. | Price per 1,000. |
|---|---|---|---|
| 1874 | 104,731,350 | 7,727,000 francs. | 73·78 francs. |
| 1875 | 227,640,212 | 11,247,416 ,, | 49·40 ,, |
| 1876 | 335,774,070 | 13,226,296 ,, | 39·39 ,, |

This shows the enormous increase of the production and the consequent lowering of the price, but not proportionately, for though the production increased more than three times the price did not fall to one-half.

The basin of Arcachon and other maritime rivers is where the artificial cultivation of oysters has mostly been attended to. In the season ending April, 1877, over 202 million, valued at $4\frac{1}{2}$ million francs, were delivered from Arcachon. The Auray district delivered over 101 million oysters, valued at 500,000 francs, during the same period.

Oyster culture in England is still in its infancy, and the oyster-beds are over-dredged and almost empty, hence the high price of the delicacy at present, but steps are being slowly taken to remedy the evil.

The following list of works on Oysters and Oyster-culture is given for those who wish to pursue the subject further :—

History of the Oyster and Oyster Fisheries ; by T. C. Eyton. London, 1858.

Die Austern und die Austernwirtshaft ; von K. Möbius. Berlin, 1877.

Die Danske Osterbanker ; H. Kroyer, Copenhagen, 1839.

On the Food and Reproductive Organs of *Ostrea virginiana*, &c. ; by J. M'Crady. Proceed. Boston Soc. Nat. History, December 3, 1873.

Unterschungen über die Austern ; K. Möbius. Nach. Deutschen Malak. Gessellschaft III, 1871.

Organes genetaux des Acephales Lamellibranches ; par H. Lacaze-Duthiers. Annales des Scienes Naturelles, 1854.

Same Author : Sur les Organes de la génération de l'Huitre. Comptes Rendus, 1855, X 4, p. 415.

On the propagation of the species in the common Oyster. Croonian Lecture for 1826, by Sir E. Horne. Phil. Trans. 1827, p. 39.

Report of the Commissioners appointed to examine into the methods of Oyster-culture in the United Kingdom and in France. Blue Book, 1870.

Guide Pratique d'Ostericulture. Prof. F. Eraiche (quoted by Dr. Brooks without date or locality).

Development of the American Oyster ; by Dr. W. K. Brooks. Annapolis, U.S. State Printer, 1880.

Report of the Oyster Culture Commission. Blue Book, Government Printing Office, Sydney, 1877.

Natural History of the Oyster and Economy of the Oyster Farm. Harvest of the Sea ; by James G. Bertram. 3rd edition. Chaps. XII and XIII. London, Murray, 1873.

Report Fisheries Inquiry Commission. Blue Book, Sydney, 1880. Evidence and Appendices *passim*.

On the Edible Oysters found on the Australian and neighbouring Coasts ; by J. E. Cox, M.D., F.L.S., &c. Proc. Linnean Society, N.S. Wales, vol. VII, p. 122.

Voyage d'Exploration sur le Littoral de la France et de l'Italie ; P. M. Coste. Imprimerie Impériale. This treats of the cultivation of the Oyster and the breeding of Eels. A magnificent book.

The Oyster ; where, how, and when to find breed, and cook it. London, Trübner & Co., 1861.

## CHAPTER VIII.

### Other Mollusca.

THE other mollusca of New South Wales are not of much importance in an economic point of view. A mussel (*Mytilus hirsutus*) is common in this harbour, and is eaten by some people. *M. latus* (var. *Dunkeri*) and *M. rostratus* are also used, but they are all dangerous articles of food at certain times.

*Trochocochlea constricta*, Lam., is used as a substitute for the British perriwinkle, but it is only consumed to a very small extent. Our common limpet, *Patella tramoserica* is quite uneatable. In speaking of the shark fishery of Tasmania, I have already referred to the animal of a large ear shell, *Haliotis nævosa*, called by fishermen the "Mutton fish," which is much esteemed by the Chinese. Mr. Chin Ateak, a Chinese merchant in Sydney, stated in his evidence before the Royal Commission that he was prepared to give 9d. a pound for it in any quantity. As the species is abundant on most parts of the coast, the collection of it might form a profitable occupation for a number of people. There are other common species on the coast.

The squids or Cephalopods are exceedingly abundant in the harbour of Port Jackson and all along the coast. They can be easily caught with a line as well as with the seine.

They might be made a source of considerable profit for exportation to Japan and China. In both of these countries all animal substances of a gelatinous character are in great request, and none more than those of the cuttle-fish tribe. The fine preparations of cuttle-fish in the Japanese Court in the Garden Palace must be fresh in the recollection of every one. It appears from the statements of Mr. Chin Ateak, that of the Cephalopods of this coast, the "Squid" (*Sepioteuthis Australis*) is highly appreciated, and in consequence highly priced. The cuttle-fish (*Sepia*) is of rather inferior quality, and the "Star-fish" of the fishermen (*Octopus*) is not used at all. Some of the latter which frequent the harbour are of enormous size. The author purchased two from fishermen which measure over 6 feet from tip to tip of the arms, each of which has over 800 suckers upon them, making considerably above 6,500 suckers for the whole fish! They are excellent eating if properly prepared, and a few words about the economical value of cuttle-fishes will here be useful. They are taken almost verbatim from Mr. Lee's excellent work on the Octopus.*

Although the cephalopods are seldom eaten in Great Britain, they are appreciated as food by nearly all other maritime nations. Along the western coast of France and in the countries bordering on the Mediterranean and Adriatic they form a portion of the habitual sustenance of the people, and are regularly exposed for sale in the markets both in a fresh and in a dried condition. Salted cuttles and octopus are eaten during Lent as commonly as salted cod is brought to table in England on Good Friday, and thus prepared generally form a portion of the

*"The Octopus," by Henry Lee, London, 1875. Chapman & Hall.

provisions supplied to the Greek fishing boats and coasters. The Indians of North-western America look upon them as the proverbial alderman regards turtle and devour them with the same gusto and relish. They roast the glutinous carcass instead of making soup of it. In Chili, Peru, Brazil, and Teneriffe they are eagerly sought for, and they are an article of daily consumption in India and China. Japan has already been mentioned.

From a Report on the Tunisian Fisheries by W. Kirby-Green, British Consul at Tunis,* we learn very interesting particulars about this fish and its fishery. The villages in the neighbourhood of Karkenah are the chief localities where the cuttle-fish are obtained, and the produce yields in a good season about 125 tons, in an average year 90 tons, in a year of scarcity 50 tons. In a good season the whole of the island of Karkenah supplies about 150 tons, and the Jerbah waters a third of this quantity. On the shores from the village of Luesa to that of Chncies, in the Gulf of Khabs, the natives collect from 4 to 5 cwt. during the season. The cuttles prefer the rocky shallows, coming from the open sea in the months of January, February, and March. The fry is observed from the months of June to August, and then the fishing is good, but if it be late, such as November and December, the following season is bad. On the arrival of the cuttles in the shallows, they keep in masses or shoals, but speedily separate in search of shelter among the rocks near the beach, covered by only 1 or 2 feet of water, and in stony localities prepared for them by the fishermen in order to facilitate the deposit of spawn. In deep water they are taken by means of earthen jars strung together and lowered to the bottom where they are allowed to remain for some time, during which the fish introduce themselves. From eight to ten cuttles are taken at each visit of the fishermen. In shallower water earthenware drain-pipes are placed side by side, for distances frequently exceeding half-a-mile in length, these the cuttles readily enter and are captured. They are attracted by all white smooth and bright substances, and the natives deck suitable places in the inlets and hollows of the rocks with white stones and shells, over which the cuttles spread themselves and are captured four to eight at a time. But the most successful manner of securing these fish is that pursued by the inhabitants of Karkenah, who form long lanes and labyrinths in the shallows by planting the butt ends of palm branches at short distances from each other, and these constructions extend over spaces of two or more miles on the ebb of the tides.† The cuttles are easily collected and strung in bunches of fifty. The capture can be increased indefinitely by the use of these labyrinths of palm branches, which seem to have a peculiar attraction for these fishes. They have been hitherto prepared for exportation by simply salting and drying; but are now preserved either in oil or brine after a previous scouring and boiling. The price varies, but the range is between 6d. and 1s. 3d. for a pair of fresh fish. Before being dried they are beaten between two stones, and this labour raises the expense from between 25s. to 50s. per cwt., to the cost price is added an export duty of 5s. 1d. per cwt. Malta

* Published in the London *Standard*, December, 1874.

† In the Syrtis the ebb and flow is 10 feet, a wonderful tide for the Mediterranean. See Richardson's "Travels in the Desert." He mentions the Palm Branch Fishery, but did not know it was for cuttle-fish.

receives the greater part of the Tunisian supplies, which are principally sold in the Greek markets. Portugal also largely competes with Tunis in supplying this trade.

Professor Forbes says :—" When well beaten to render the flesh tender before being dressed, and then cut up into morsels and served in a savoury brown stew, they make a dish by no means to be despised, excellent in both substance and flavour. A modern Lycian dinner, in which stewed cuttle-fish formed the first and roast porcupine the second course, would scarcely fail to be relished by an unprejudiced epicure in search of novelty."* Mr. Lee says that they may also be eaten plainly boiled and served with egg sauce. The flavour is not unlike that of the skate or the soft part of the scallop.

More than a hundred millions of cod are caught annually with cuttle-fish as bait.† This bait is used in August and September, for they then come into the fishing-ground in abundance. They are caught by means of a jigger or conical piece of lead, round the base of which eight or ten hooks are inserted. The fishermen go out in punts "squid-jigging" of an evening to catch the bait required for the next day's fishing. Whilst they fish they shout and make a great noise, for some unknown reason. All parts of the squid are used as bait, but they sometimes are eaten by the fishermen, who say that they are remarkably sweet and excellent when fried. The eyes of the cephalopods are very solid, formed of two double concave lenses, separated by a groove. The two halves easily separate, and exhibit internally a series of concentric coats of beautiful nacreous aspect. They are used as ornaments in Italy, and sold as pearls in the Sandwich Islands. The ink which is ejected by squids has been used as a pigment from classical times. At present in London thousands of "ink-bags" are manufactured into "sepia" by artists' colourmen. The fishermen of some of the English northern counties, when cleaning cuttle-fish for bait, dry the ink-bags and their contents, and sell them to the colour-manufacturers.

* " Travels in Lycia," by Spratt and Forbes.
† Edin., New Phil. Jour., 1826, p. 32.

# CHAPTER IX.
## Crustacea.

The Crustacea of Australia have recently formed the subject of special study by Mr. W. A. Haswell, and the results have been published by the Trustees of the Australian Museum.* From this work it appears that our species are numerous, though very few are used as articles of food.

## The Sea Crab.
(Plate XLII.)

Two swimming crabs are often seen in the market—*Scylla serrata* and *Neptunus pelagicus*, Linn† They both have a very wide range. *S. serrata* is taken in the Endeavour River as well as at Port Jackson. They are generally taken in the seine-net, and being of excellent quality are readily bought up in the market. With one exception, however, they are too small to be of much use, the exception being *Scylla serrata*, which is said to attain a weight of 4 lbs. and to be equal to the Scotch partan. It is however seldom caught, though there is no reason to believe that it is rare on our coasts. Nothing seems to have been observed as to the habits and breeding-time of those crustacea. None of the numerous shore crabs are ever used as food. One species of the latter is very common all along the east coast, from the extreme north to Port Jackson, *i.e. Grapsus variegatus*, Fab. This species extends all through the Pacific to the coasts of Chili.

## The Cray-fish.
(Plate XLIII.)

The Marcourous Crustacea are better known. The large cray-fish (*Palinurus hugelii*, Heller) is about the finest of its kind. It attains a weight of 6 lbs., and when in season is entirely filled with meat of the most excellent quality. It is found during the summer season abundantly in most of our bays and along all parts of the coast, and is caught by means of circular hand-nets from boats. In the early part of summer the ova are found within the abdomen of the female, when it is known to gourmets as the "coral." The animal is then in the finest condition. About January the ova are shed, and are carried for a long time in a semi-hatched state on the underside of the tail or abdomen of the mother. Unfortunately the quality for food of the cray-fish does not seem to suffer much at this period, as is the case with crabs and other crustacea while undergoing the above process; the consequence is that the destruction of the young fish with the mother is going on during the entire season. The consumption in Sydney of this crustacean is very considerable, and the price is generally high, but at times there is a glut in the market, and in consequence a serious fall in price. There are many parts of the coast too distant from Sydney for the supply of that market, where we believe that establishments for the "canning" of this most valuable crustacean could be most profitably undertaken. One such place is the Broughton Islands, a few miles north of Port Stephens, where the supply of this crustacean is unlimited. The demand for tinned lobster is great all over the world, and in North America, from which it chiefly comes, the supply is rapidly diminishing.

* "The Stalk and Sessile-eyed Crustacea of Australia," by W. A. Haswell, Sydney, 1882.

† This crab is very widely distributed, and swims in the deepest seas.

The other species of cray-fish *(Palinurus lalandii)* is often to be seen in the shop-windows of the Sydney fishmongers, but it chiefly comes from Tasmania. It is seldom found on the coast of New South Wales north of Twofold Bay. It is not nearly so large as the Sydney cray-fish, but is said to be equally valuable as food.
—R.R.C.

## The Prawn.

The prawn *(Penæus esculentus)* is abundant in most of our shallow bays and harbours, and is a most popular article of food amongst all classes. The consumption of this crustacean is so great that fears have been expressed that the supply might become exhausted, and it is undoubted that the size of those brought to market now is often much below the average of former years. We do not believe, however, that there is any danger of exhausting the supply for a long time to come, and the only limit we would suggest to the capture of prawns is what we have already advised for the protection of young fish generally—the limitation of the length of the prawn nets and the size of the mesh.—R.R.C.

There are two well-known species of *Penæus*, the one just mentioned, and *P. macleayi*. The first reaches a length of 9 inches, and the second is not quite as large. *Alphæus socialis*, Heller, locally named the "Nipper," is abundant in Port Jackson, and is a good deal sought for, but not so much for food as for bait for black bream fishing. All these prawns go a certain distance up the freshwater rivers. In the Hunter, about West Maitland, they are much used in the end of summer as a bait for perch (*Lates*), bream, and flathead.

We have also a kind of prawn in all the western fresh-water rivers. It has long narrow claws and attains a pretty good size, so that it is used as an article of food. This is described as *Palæmon ornatus*. The only true shrimp (*Crangon*) which Australian waters are known to possess is found in the Gulf of S. Vincent, S. Australia.

## The Freshwater Cray-fish.

(Plate XLIV.)

1. Side view, one-third natural size. 1*a*. Rostrum, natural size. 1*b*. Claw, natural size, showing peculiar spines. 1*c*. Tail.

We have several species of fresh-water cray-fish in Australia, belonging to the genus *Astacopsis*, or as some naturalists call it *Astacoides*. The genus *Astacopsis* differs from *Astacus*, the common European fresh-water cray-fish, principally in having no appendices to the first segment of the abdomen and some minor peculiarities of the feet. *A. bicarinatus* is abundant in all waterholes of most of the continent, Port Essington, Cape York, Brisbane, and in both eastern and western waters. These cray-fish prefer still waters, and burrow in the banks, often doing much injury to dams, &c. They can brave long droughts in their burrows. In the western waters of New South Wales we have a larger and much more ornamental species, *A. serratus*. (Plate XLIV). *A. plebius* is attributed to Sydney, but it is probably from the Pacific Islands. There are different species in Tasmania (*A. franklinii*), West Australia (*A. 5-carinatus, preissii*), and North Australia (*A. 4-carinatus*). There is also another species in the swamps of the extreme west, which may be another species of the *Engæus* of Tasmania, which is a small cray-fish living in burrows in the swamps, considered by most naturalists a sub-genus of *Astacopsis*. *Engæus fossor* is found in Gippsland : a species of small size and no value, though the natives used to consume them.

It is computed that our common saltwater cray-fish produces 100,000 eggs each season. The spawn can be artificially hatched with the greatest ease. A series of pans with sand and gravel is all that is required. The eggs can be laid on these and water be allowed to flow over them. The water must be salt or fresh according as the cray-fish is marine or not. In rearing the young crustaceans it must be remembered that their food in a state of nature consists of marine worms, fish spawn, and other small crustaceans, and this sort of food must be provided. In France cray-fish hatching has been very successfully tried.

Of the true lobsters New South Wales can only boast of a very few, and these entirely confined to fresh water. Under the name "Marami" are included two or more species of *Astacopsis*, found in all the creeks and mud-holes of the country. They are not much used as food, probably on account of their small size—seldom exceeding 4 or 5 inches in length, for they are as good to eat as any of the tribe. One species, however, forms an exception to the others in point of size; it is the lobster of the Murrumbidgee and Murray system of Rivers—*Astacopsis serratus*. This beautifully coloured lobster attains a considerable size, averaging a foot in length, and is esteemed a great delicacy. It is largely consumed by the residents on these rivers during the winter months—the season when they are in the best condition and most readily caught.—R.R.C.

A very good figure of *A. bicarinatus* is given by Prof. M'Coy, in the Prodromus Zool. Victoria, pl. 29. It differs from the species figured in our Plate XLIV, in being destitute of those spines and tubercles which ornament the shell of *A. serratus*, and it is smaller. It grows to about 6 inches in length, but varies much in colour, ranging from yellow to brown or horn colour and dark olive. The claws are always blue with red joints in living specimens, but the smaller legs are blue, or green, or whitish. They used to be prized as food by the aborigines; and near the swamps and rivers of Victoria heaps of their remains may be seen in the old "middens" of the natives. Some white people like them too, but the flavour is decidedly muddy. I have reason to be grateful to them. In 1856, in a long overland journey between Victoria and South Australia, I must have suffered great exhaustion but for the food these cray-fishes afforded me.

## CHAPTER X.

### The Fishing-grounds of New South Wales.

THE following facts with reference to our Fishing-grounds have been elicited by the evidence obtained by the Royal Commission of 1880, and were embodied in their report. They comprise all the information we have on the subject.

The number, variety, and extent of the fishing-grounds with which the entire seaboard of the Colony has been endowed, all lying within a very moderate distance of Port Jackson, afford the strongest encouragement to the inhabitants of the whole Colony who may hope to see the markets supplied with fish in a manner and upon a system consonant with the requirements of the community. In this Colony the fish most adapted for food purposes do not yet require to be searched for in large smacks or fishing-vessels, victualled and equipped for a cruise of several months; neither is it necessary for our fishermen to make voyages to fishing-grounds distant hundreds of miles from home. Our best fish are very rarely met with more than 10 miles off the coast, or in deeper water than 35 fathoms. The schnapper, which for economic purposes may be ranked with the cod of the northern hemisphere, appears to be distributed with remarkable regularity along the whole extent of our seaboard—that is to say, over about 600 miles; and whatever the formation or character of the coast may be, this fish, the most valuable of all our forms, and perhaps the most abundant, is never absent; and being essentially a rock fish in its habits, is not migratory. And the same may be said of its congener—the bream; and in a lesser degree of the flathead, whiting, black-fish, tailors, tarwhine, garfish, and other varieties which frequent the bays and estuaries of harbours and lakes, rather than the ocean depths. Some of these fish are, no doubt, not to be found throughout the year in their usual haunts, but they may be treated for all practical purposes as regular inhabitants of our fishing grounds.

The seaboard of this Colony is in a marked degree favourable to the existence of a very large supply of the best food fishes. It is indented by innumerable inlets and arms of the sea; it possesses many rivers whose embouchures are of large expanse; some of its bays, harbours, and lakes are of vast extent, and its submarine conditions generally are of a character eminently adapted both as nursery and feeding-grounds for fish.

Port Jackson, although not very many years ago holding a very high rank among our fishing-grounds for all kinds of the best net fish, is now scarcely regarded as a source of supply at all. And this is owing not so much to the pollution of its waters by the sewage of a large city, or their constant disturbance by the traffic of innumerable vessels, as to the ceaseless and often wanton process of netting to which every bay and flat has been subjected for the past fifteen or twenty years. The wholesale destruction within the harbour caused by stake nets and seines with meshes almost small enough for a naturalist's hand-net has of course produced its natural effect on the outside grounds, where the schnapper can now only be taken in very small quantities, and without any degree of certainty. The evidence given by fishermen who can remember the large hauls of fish once taken from the beaches of North and Middle Harbour, Rose and Double Bay, not to speak of the flats up the Parramatta River, affirms this.

The fisheries of New South Wales were classified into three large groups by the Report of the Royal Commission, which described them as follows :—

(1.) The Home grounds—comprising those lying between Cape Three Points and Sydney to the North, and between Sydney and Wattamolle to the south. The most remote grounds between these limits represent tolerably well the terminal points for all open fishing-boats during the summer months.

(2.) The Middle grounds—which to the north would be comprised between Cape Three Points and the inlet known as Cape Hawke—and to the South would include those lying between Wattamolle and Wreck Bite. These represent the present limits both of open-boat fishing and of supply by steam coasters during the winter months.

(3.) The Outer grounds, which would embrace the grounds north of Cape Hawke and south of Wreck Bight as far as the northern and southern boundaries of the Colony, or say as far as the Tweed River to the north and Twofold Bay to the south. These grounds are too far distant from a market to be available for the supply of fresh fish, until at least some such fishing-vessels as those recommended later on in our Report shall engage in the trade.

(1.) *The Home Grounds.*—Within this section the embouchure and lower waters of the river Hawkesbury, better known as Broken Bay, situated about 16 miles from Port Jackson Heads, has always ranked, and perhaps ranks still, as the most extensive and most productive of all our fishing stations. The beaches of Pittwater, the Hawkesbury proper, and Brisbane Water, present the most favourable conditions for the net fisherman, and the upper reaches of the river and the mud flats of its various tributaries, especially at the places locally known as Mullet Island Creek, Mooney, Mother Marr, Berowa, and Mangrove, have supplied to the Sydney market for many years past, and may under proper restrictions and protection long continue to supply, enormous freights of the choicest of our river fishes, such as black and sea bream, tarwhine, black-fish, whiting, flathead, tailors, garfish, and the large sea and flat-tailed mullet.

Equally prolific have the outer or schnapper grounds at and near the mouth of this river been to the line fisherman. These are to be found in great variety from Cape Three Points to the South Head of Broken Bay. Off the North Head of the Bay, and again off Little Head, situated a few miles to the southward, there occur several schnapper grounds of high renown, which a few years ago kept as many as a dozen or more boats in full work for the Chinese fish-curers, who were then engaged in a large business on Schnapper-man's Flat, Pittwater, but who have now entirely abandoned Broken Bay as a fishing station. Twenty and thirty dozen "count" fish (*i.e.*, fish weighing each 6 lbs. or over), were often taken by two fishermen on these grounds. Now, however, both the schnapper and the net grounds about Broken Bay have fallen off in their productiveness to an alarming degree, and through the operation of the same destructive agencies which have brought about the impoverishment of the Port Jackson and adjacent fishing-grounds.

Being however quite out of the beat of any constable and not readily accessible except by water, and possessing in other respects features favourable for the prosecution of net-fishing by fishermen not always over-scrupulous in obeying the directions of our laws with reference to size of mesh or other statutory restraints, the Broken Bay fishing-grounds are still much resorted to. At the present time the number of boats frequenting them is estimated by a competent witness at about twenty-five on an average of the year.

Many freights of valuable fish are thrown away or abandoned and left to rot on the beaches about Broken Bay during the summer months, and not unfrequently in the colder months. This happens when the fisherman coming down the river with his boat fully laden finds a strong southerly wind blowing in the offing. He has no means of taking his fish to market overland, although the distance from the head of Pittwater to Manly cannot be more than about 12 miles. Accordingly, unless the head wind takes off or veers to a favourable quarter within twenty-four hours at most, he cannot possibly expect to save his freight, unless he have the good luck to catch the steamer which makes occasional visits to Broken Bay.

If the line of the proposed Northern Railway extension to Sydney should happen to cross the Hawkesbury at any point within 10 miles of Broken Bay, enormous losses of fish by stress of weather will be prevented, as the fish caught in the lower waters of the river or on the outside grounds may then be easily and rapidly transported to market by rail.

The most important fishing-grounds lying between Broken Bay and the southern limit which we have given to the home fisheries are, after leaving Little Head, the schnapper grounds, situated at varying distances from all the main headlands, such as the South Head of Broken Bay, Bungy, Little Reef, Narrabeen, Long Reef, Deewhy, Curl Curl, and North Head (Port Jackson). There are at least a score of known "school fish" grounds within these limits, *i.e.*, about 12 miles. Vast quantities of fish have been taken on every one of them, especially along the line of submerged rocks known as Long Reef, which can be traced for a distance of more than 5 miles from the shore, and on the wide grounds off Narrabeen. Narrabeen is the bight or bay which lies immediately to the north of Long Reef Point, and marks the small and rarely navigable sea mouth of the lagoon known by the same name. Other lagoons, such as Deewhy and Curl Curl are found in this locality, but none are of such extent or importance as Narrabeen which in times past has been a very productive netting ground. A few "bumboras" are found in this bight, and they (like all "bumboras" on the coast) have been and still are the favourite resort of the schnapper-men during particular conditions of the currents.

The Long Reef grounds extend for a considerable distance seaward, and being distant only some 5 miles from Sydney Heads, these grounds have perhaps been more fished than any in the vicinity of Sydney, and have, it must be confessed, stood the strain on their resources in a remarkable manner.

The ordinary varieties of fish seem most to frequent the feeding-grounds presented by this extensive backbone of rocks during the prevalence of southerly currents, and especially after light easterly weather. Fish are here most abundant during the summer months. At present the fishermen complain bitterly of the leather-jackets which infest these and other grounds near Sydney, and it is no doubt true that these execrable pests have been gradually increasing in numbers. Their habit is to lie between the surface and the bottom where the schnappers feed, and to gnaw off every bait, often with the snood too, as it descends. Their numbers do not seem to diminish, however many are caught, and the damage done in a single day to a fisherman's gear has not seldom reached a far higher sum than the value of his freight of marketable fish. As food this fish is not appreciated, though it is by no means to be despised. The leather-jacket, however, has not the appearance of a food fish, and in this respect, like the cat-fish, owes much of his immunity to a rather repulsive exterior.

It is said that some wide grounds have lately been found at a distance of 10 miles or more from Long Reef Point, where in about 35 to 40 fathoms of water schnappers have been taken in large quantities. Between Long Reef and the North Head the bottom is almost uniformly foul, and no good outer grounds excepting those off Deewhy are found hereabout, at least for school fish. One may drop accidentally upon a few fish almost anywhere, but these constantly recurring patches of rocks appear to be barren of the small crustacea and other food affected by schnappers, until we reach the line of reefs jutting out (under water) from North Head. This, like Long Reef, forms a series of fishing-grounds for schnappers, with a reputation of long standing, but they are not now much to be depended on. Occasionally a few boats may be seen (chiefly of the amateur fishing class) off these grounds, and on the Pine-tree, the Cobbler's ground, and others in the vicinity, but the fish no longer frequent these places in payable quantities. There was once a good fishing-ground lying about due east, and at a distance of about 3 miles from the Heads, but it is said that the mud-punts which here discharge the silt and harbour filth have quite disgusted the schnapper, whatever attractions they may have created for the "leather-jackets."

The next grounds resorted to by fishermen are off the Flagstaff, and off Mud Island (near the proposed outlet of the main sewer), and then passing a few others of less note we come to the celebrated Coogee fishing-grounds. These are situated at varying distances of 1 to 3 miles from a small rocky islet known as the "Island of Rocks," and lying a short distance from the shore of Coogee Bay. Enormous freights of schnapper have been taken along this line of grounds, and fair quantities are still taken. This is a favourite fishing-ground for the various fishing clubs which pursue their recreation in steam launches and steam tugs, and by sheer force of numbers and the ease with which they are able to shift from one spot to another without any picking up of heavy kellicks or beating to windward, are able to count out more fish on a fair day's "outing" than any of our professional fishing crews. A thousand fish, we are informed, is not an extraordinary catch for some of these clubs. While on the subject of fishing clubs we may here mention that a very intelligent witness examined by the Commission (Mr. M'Carthy) has informed us of the discovery by him of an extensive "shell bank," at a distance of about 10 miles eastward of these Coogee grounds. This bank is said to carry only about 20 to 30 fathoms of water, and to be very narrow, in fact so narrow that the steamer finds great difficulty in lying-to on it. The soundings in the immediate neighbourhood of the ridge, which extends in a N.W. and S.E. direction for about a mile (so far as Mr. M'Carthy was able to trace it), show from 70 to 90 fathoms, from which we are able to form some idea of the character of this remarkable upheaval or submarine sand-ridge.

Between Coogee and Cape Banks (the northern headland of Botany Bay) the fishing-grounds are wholly confined to those in the offing, there being no lagoons or inlets capable of being worked by nets. There are about a dozen schnapper grounds within these limits, but none of them are of much importance. Nannygey are often caught in quantities in the depths off Long Bay Head, and schnapper have at times been found to abound off Marubera and Little Bay (wide); but until we reach the long line of rocky ground which forms the submarine extension of Cape Banks, none of the grounds hereabout possess the necessary conditions for school fish. What fish there are do not seem to be settled on any one ground, but, as is the case on all foul grounds, they roam from patch to patch in small schools. The entrance to Botany is foul ground as a rule, and although wide off Cape Banks there are some very fair school-fish grounds, yet they have never been much appreciated by fishermen, who prefer the "bumboras" and off shore grounds to the southward of Cape Solander, Long Nose, and Curranulla Head, notwithstanding the strength of the southerly current in the summer months which, off some of these headlands, runs like a sluice. But Botany, though never equal to the Long Reef and Broken Bay grounds for school-fish, has always held its own for net-fish ; indeed it is doubtful whether even Broken Bay, with its far greater extent of net grounds, has ever been or is now more productive than are the beaches and flats of Botany. This inlet, though shallow, covers many thousand acres of water ; and as two salt-water rivers—George's and Cook's—flow into it, there is no lack of spawning or nursery grounds. Although of late years as many as a dozen net boats have regularly plied the beaches and flats of Broken Bay, yet fish are still caught. Of course the processes of exhaustion common to all the grounds near Sydney have been in active operation at Botany, and especially that most destructive of all forms of fishery the stake-net ; yet notwithstanding all this, Botany is perhaps less impoverished considering the amount of fishing continually going on than any inlet or harbour within fisherman's distance of Sydney. But the evidence of witnesses well acquainted with the resources of Botany leads immediately to the conclusion that, unless arrested by legislative restraints, these prolific grounds will in a very short time succumb to the stake-nets and the small mesh as surely as our other fishing-grounds have succumbed.

FISH AND FISHERIES. 133

The following extract from Mr. Oliver's paper on the Fisheries of the Colony describes the grounds from Botany to Wattamolle :—" At and off the entrance of Botany and Curranulla Head there are several well known schnapper grounds, and about two miles within Curranulla bight (the "Bate Bay" of our charts) is a famous ground known to fishermen as the Mary, Merry, or Shamrock Rock, for it goes under all these names. It is a sunken flat rock, or series of rocks, with about 8 to 11 fathoms of water, situated at the point of a reef which runs from a little boat-harbour called 'Doughboy,' about half a mile to the southward. Tons upon tons of schnappers have been taken off this ground, which however is difficult for a stranger to find, as the crossbearing marks are not easily described. The whole of this Port Hacking or Curranulla Bight is one vast nursery and feeding-ground for fish, and the harbour and river of Port Hacking at its southern extremity is second only to Broken Bay as a net ground. Here are caught generally the first garfish and mullet of the season, both which fish come to us from the southward, generally seeking the smooth harbour waters after heavy south and south-easterly weather, and, after a few days continuing their progress northward, and putting in at every inlet or river-mouth lying in their course. A cable-length or so distant from 'Jibben Head,' the southern point of the entrance to Port Hacking, lies Jibben 'bumbora,' a fishing-mark of great repute, but not now much resorted to for school-fish, i.e., the schnapper of about four to six or seven years old and found on the off-shore grounds in large schools, as distinguished from the native, which is the same fish at a later stage of growth, but frequenting different haunts (the shoals off headlands, sunken rocks, and river points. Passing south, the inshore grounds off Marly Head and Wattamolle are next reached, and this latter point forms the Sydney and Botany fishermen's *Ultima Thule*. Indeed, these southern fishing-grounds are rarely troubled, except in the winter months, when the wind generally blows off the shore, and is fair for both the up and the down trip."

(2.) *The Middle Grounds (North).*—Between Cape Three Points, a few miles to the northward of Broken Bay, and the inlet commonly known as Cape Hawke, a section comprising about 100 miles of coast-line, both the breeding and offing grounds are perhaps more abundant and of greater extent than are to be found anywhere on our coast. In the first place we reach, soon after passing Terrigal High Land—Point Upright of the charts—the famous Tuggerah Lakes. This series of lakes or salt-water lagoons, consists of Tuggerah (proper) and two smaller connected lakes, locally known as Budgewi and Mánnura. They are all comparatively shallow, perhaps averaging rather less than 2 fathoms, though in some places, *e.g.*, near Woollahra Point in Lake Tuggerah, the depth is considerably more. Tuggerah is by far the largest of the series, being about 9 miles long by about $2\frac{1}{2}$ in width. The others are much smaller.

The sea connection with Tuggerah is at a small rocky opening in the beach about 3 miles north of Point Upright, and about 7 from the well-known boat harbour called Terrigal. A dangerous line of reefs runs out in a line about E.S.E. from the entrance, terminating in one of the most treacherous "bumboras" on the coast. This entrance is unfortunately a very broken and shallow one, and is rarely available even for open boats; still, during the prevalence of westerly winds fishermen's boats do occasionally manage to navigate it.

A few miles to the northward of Manmura commences the still larger expanse of waters known as Lake Macquarie. This lake is nearly 20 miles in length, by an average of about 3 miles in width, but its contour is so broken by deeply indented bays and recesses as to give a perimeter of about 300 miles. Unlike Tuggerah, Lake Macquarie possesses a very tolerable entrance, available for craft drawing up to 6 feet of water, and when the works now in progress in the river channel and at the entrance are completed Lake Macquarie will probably be navigable for vessels of 10 feet draught. The average depth of Lake Macquarie is about the same as that of Tuggerah. These lakes are the great nurseries of almost all our winter supplies of net and line fish. Here unquestionably the sea mullet, bream, tarwhine, whiting, flathead, tailor, and garfish find their most congenial spawning-grounds, and here also are their natural sanctuaries from sharks and other predaceous fishes which devour them in the offing. Here also, it is believed, is the chief spawning-ground of the schnapper, which afterwards haunts the numerous reefs, bumboras, and rocky patches which lie between Broken Bay and Newcastle. The supplies of fish from the Tuggerah Lakes have during the past few years been despatched to the metropolis by a small freight steamer, which goes to Bungaree or Bungaree's Norah (the Norah Head of the charts), a long low point which forms the northern extremity of Tuggerah Bight. Inside the reef and the low-water ledge of rocks, which form a kind of breakwater, is a tolerable anchorage for small craft and boats, and a not very good landing, whence, however, are shipped many fine freights of the best fish which come to our market. From Terrigal to Bird Island the offing and inshore grounds still abound in all the best kinds of line fish. It is almost impossible to find a square furlong untenanted by the schnapper or other equally good fish. The fish are, however, at times in the habit of shifting their quarters from one ground to another in the neighbourhood. These grounds are mostly fished by schnapper men, who camp at Terrigal or Norah, and chiefly in the cool months of the year, the distance from market (over 30 miles) proving a formidable obstacle to even the most hardy and enterprising fishermen during the prevalence of strong southerly or north-easterly winds. With proper freight steamers of course this obstacle will speedily disappear, and we shall then get fresh schnappers and black rock cod (which are here caught of great size and excellence) from the Tuggerah and Norah bumboras and the Bird Island grounds, with as much regularity, but in far greater quantities, than we are now able to furnish ourselves with from Long Reef or Coogee. At present we are informed that there are no Chinamen on the Tuggerah Lakes, and only a few at Lake Macquarie. This is undoubtedly not a subject for regret, although during last winter their places have been filled by men whose engines of destruction were far more formidable than the Chinese. We have been told of fixed nets being used in Tuggerah Lake enclosing an area of more than half a square mile of water, having meshes so small that nothing could escape. This wholesale process—rather facetiously known as "mortgaging"—was supplemented by an auxiliary proceeding (presumably of "foreclosure"). Inside the huge ring made by a mile or so of net, a boat from time to time throws off a small seine, which is "bull-ringed," or drawn to the shore where practicable round as many fish as are required for the next trip of the steamer. We have been told on good authority that the proceeds of sale averaged for a long time about £100 per month to one fisherman; very handsome results, considering the outrageous violation of the law by which they were procured.

Crayfish are known to abound off some of the headlands in this neighbourhood, but at present no supplies seem to come from these waters to Sydney. From Bird Island to Nobby's (Newcastle) schnapper grounds are numerous, but not so abundant, or, it may be, not so well known as those about Tuggerah Bight and Bird Island. Off Wabung and Spoon Island reefs the inshore grounds are well furnished; and again, those lying off the entrance to Lake Macquarie ("Reid's Mistake" of the charts).

In the bight between the small island so known (it is called "Creen" Island by the aborigines and "Green" Island by the coasters and fishermen) and Red Head, there are several "bumboras" and rocky patches where schnappers can be taken in almost any quantities, but the sharks are usually very plentiful also in these localities.

Fresh fish are not often taken to Sydney from Lake Macquarie, the few fishermen stationed there preferring to fish for the Chinese curers rather than to take the chance of catching a Newcastle steamer four or five miles in the offing. Large quantities of mullet were at one time cured here for the Newcastle market, and it is said that a considerable quantity of fresh fish finds its way to the Wallsend mining population at the north end of the lake. We cannot leave this "Lake" section of our northern fishing-grounds, as it might very aptly be termed, without expressing an emphatic opinion as to the urgent necessity of protecting by some effective legislation these magnificent "nurseries" from any further destruction by nets of unlimited length and diminutive mesh, such as an eye-witness has told us have at one haul frequently brought to shore a *ton* or more of small fish, for no better purpose than to be left to rot there. In the economy of our Fisheries these warm and sheltered waters, abounding as they do in minute crustacea and other food, play a most important part; and if those in the neighbourhood of our large centres of population be not soon relieved from the wantonly destructive agencies which are now ruining the young fry, of which these lakes are the natural homes, it will be futile to expect any considerable results from the protection, at spawning-time, of adult fish—at all events within the range of waters for which these inlets are the appointed nurseries.

Newcastle and the lower reaches of the river Hunter are at present of far more importance to Sydney as the chief station of the prawn fishery, and for their natural and other oysters beds, than for the supply of line or net fish which they afford. It is even said that the population of Newcastle is not adequately supplied by the Hunter; and the great and constant destruction of small fish by prawn nets is stated to be the not improbable reason. Port Stephens, about 24 miles to the northward of Newcastle, with its innumerable outer grounds, including the Broughton Islands, and extending as far as the Seal Rocks, is probably the grandest fishing station on the entire seaboard of this Colony. Connected with the vast series of lakes (the Myall Lakes) on the north, and with the Karuah River, Telligherry Creek, and half a score of important affluents inland, with miles upon miles of beaches fit for seine fishing, with an apparently unlimited endowment of the best fish, and with a telegraph station within very easy distance, this noble harbour is unquestionably destined to become one of the largest factors in the metropolitan fish supply of the future. A considerable gang of Chinamen is always located at Nelson Bay, and as soon as one lot returns to its native country another takes its place. They catch their own fish here, and preserve it after their own detestable fashion. At times these Chinese fishermen go out seaward as far as Long Island, where they camp for several weeks at a time and catch vast quantities of fish, and might, if they chose, catch any quantity of crayfish (the "lobsters" of our fishermen). All the Broughton Island group are singularly favourable by their formation for these crustaceans, and the reefs and outlying rocks about Cabbage-tree and Boondelbah Islands, off the entrance of Port Stephens, and those which lie around Long Island, abound with this fish; but the distance, under the existing *régime*, is rather too remote from market to encourage the development of this fishery.

Some 15 miles to the north-east of Long Island are the Seal Rocks, and a variety of reefs and rocky patches, all lying within a range of a few miles from Sugar-loaf Point, including the "bumbora" known on the charts by the name of the "Edith Breaker." This is a great country for schnapper and black rock cod. Sharks, unfortunately, are rather too plentiful; but there are few fishermen who venture so far from Port Stephens as the Seal Rocks, and were it not for the coasters who are sometimes becalmed hereabout, and for occasional visitations of the Marine Board, little would be known of these grounds. Between Sugar-loaf Point and Cape Hawke, another 15 miles further north, there is no lack of virgin schnapper grounds, and a few miles beyond Cape Hawke is the outlet of the Wallis Lakes,—another series of rather shallow but extensive lagoons, similar in general character to those which we have already noticed. These lakes teem with the best of net fish, but owing to their distance from market enjoy a complete immunity from all sources of disturbance except the occasional net of some settler or sawyer.

*The Middle Grounds (South)*.—This section includes the coast between Wattamolle—a small boat harbour about 5 miles to the south of Port Hacking—and a rather deep indent immediately to the southward of Jervis Bay, known as Wreck Bay or Bight. It comprises about the same length of coast line as the northern portion of the Middle Grounds section, but, with the exception of Lake Illawarra, the Shoalhaven and Crookhaven Rivers, Jervis Bay, and St. George's Basin, possesses few inlets or sheltered waters. This deficiency is, however, fully compensated by the magnificent fishing-grounds of Jervis Bay and the extensive banks lying toward the south-eastern corner of Shoalhaven Bight, and known as Young Banks. About Stanfield Bay, to the south of the Wattamolle boat harbour, some very good schnapper grounds afford fine freights in the winter months, but beyond this point no line fish come to the Sydney Market, the distance from Sydney being too great for the class of fishing-boat employed. Lake Illawarra is fished by nets for the Sydney Market with considerable success, the fish being brought to port by steamer during the cool season of the year. Schnappers are said to be abundant off the Five Islands and other grounds in this neighbourhood, and doubtless a little search would discover many such grounds between Point Bass and Shoalhaven Bight. The waters of the Shoalhaven and Crookhaven abound in the best kind of harbour fish, and the whiting are specially excellent; but here again the want of speedy conveyance to market operates as an effective prohibition against all enterprise. The banks just mentioned cannot, however, long remain in their present state of isolation, but must before long be laid under contribution as a source of supply. The same remark applies to Jervis Bay in respect to net fish. Its beaches in extent and productiveness are probably unsurpassed by any harbour or inlet on the whole of our coast. Whiting are caught in the bay and on the outer beaches of Wreck Bight in enormous quantities, and were a short time ago, if they are not still, preserved in a dried or salted state by the Chinese fishermen. St. George's Basin, at the southern end of this bight, is a vast nursery for mullet, river garfish, bream, and the ordinary kinds of net fish, but very few are ever taken. Jervis Bay, like Port Stephens, is connected with Sydney by telegraph, and like Port Stephens is a safe and convenient harbour in any weather. Each of these stations is distant from Sydney not more than a few hours by steam, and the facilities for catching large freights of every kind of marketable fish, as well as for curing and utilizing ashore such as may be either unsuitable to or in excess of the demand are common to both. The telegraph, it is almost needless to point out, would be an invaluable guide as to the state of the metropolitan supply. Here therefore, and at Port Stephens, we should expect, in the event of our recommendations for the alternate closing of some of the home fisheries being adopted, that a large fishing industry will be established and worked in connection with steam fishing-vessels of the class.

(3.) *The Outer Grounds.*—This section of our sources of supply embraces the remainder of our fishing territory, from Cape Hawke to the Tweed on the north—from Wreck Bight to Twofold Bay on the south. As might be anticipated from the remoteness of the fishing-grounds comprised within these limits, they are almost untouched, and indeed almost unknown, at all events so far as professional fishermen are concerned.

The only information about them which we possess comes chiefly from coasting mariners and from the pilots or other officers stationed at the mouths of the various bar harbours which mark the outlets of such rivers as the Tweed, the Richmond and Clarence, the Nambuccra, Bellinger, Hastings, Macleay, Manning, Clyde, Moruya; also at Twofold Bay, and other points of less importance. From other sources we have been furnished with descriptions of such fishing-grounds as our informants happened to be familiar with; and some of your Commissioners have themselves been able, from personal knowledge of the localities, to confirm or supplement the information so obtained.

The Solitaries and other rocky islets lying between Trial Bay and the mouth of the Clarence enjoy a high reputation for schnappers and other valuable kinds of line fish; and the lower waters of the various rivers which enter the sea within this section of our fishing-grounds are known to abound, like all other rivers on the coast, with the best descriptions of net fish. Similar accounts are given of the outer grounds to the southward. In short, it may be said with perfect truth, that the whole of these portions of our coast waters are plentifully supplied with fish of

s

the same species which are found on the home and middle grounds. The time is not far distant when these outer grounds must be laid under contribution, but for the next few years we are of opinion that the supplies of fish procurable on the less remote and more easily accessible grounds will suffice for all our requirements, not only in the way of fresh fish, but also of dried or preserved fish. Still, if suitable vessels were constructed and equipped, and if the necessary capital and enterprise were forthcoming, there can be no doubt that a profitable industry lies ready for immediate development, not only on the middle, but also on the outer fisheries. The demands of the metropolis and inland towns may, however, for some time be served by the establishment of properly organized fishing stations within a distance of 100 miles north and south of Port Jackson. We have already drawn attention to the very favourable situation and conditions of Port Stephens and Jervis Bay for this purpose, and we are of opinion that better positions for fishing villages and for the establishment of industries connected with the curing or other utilization of surplus catches cannot be found on our coast.

Before leaving this portion of our subject, we desire to draw attention to the alleged existence of extensive shell-banks or submarine ridges in the immediate vicinity of Port Jackson. We are informed and believe that these banks are a favourite resort for schnapper and other deep-sea fish; and it is not at all improbable that similar banks would be discovered along the coast if a systematic search were entered upon.

The following are the distances of the fishing grounds from Sydney :—

*To the North.*

| | |
|---|---|
| Broken Bay | 20 miles. |
| Tuggerah Beach Lake | 40 ,, |
| Lake Macquarie | 55 ,, |
| Port Stephens | 75 ,, |
| Cape Hawke | 113 ,, |
| River Manning | 150 ,, |

And the Seal Rocks, 100 miles, for schnapper fishing.

*To the South.*

| | |
|---|---|
| Shoalhaven | 75 miles. |
| Jervis Bay | 100 ,, |
| Sir George's Basin | 115 ,, |
| Clyde River | 150 ,, |

All for net fishing.

There are other fishing grounds further away both to the north and south, but the distance would probably be considered too great.

*River Fisheries.*—I add here a few remarks on the river fisheries, which was not included in the subject as examined by the Commissioners. The differences between the freshwater fishes are dealt with in the chapter devoted to them, where the peculiarities of the eastern and western waters are described. In the eastern rivers there is no fishery, properly speaking; the freshwater kinds are only sought by anglers and sportsmen. The list generally includes perch (*Lates colonorum*), a herring (*Clupea novæ-hollandiæ*), which is also called a sprat, bream (*Chrysophrys*), a second kind of perch (*Anthias ?*), two eels (*Anguilla australis* and *Murænesox cinereus*), two or three kinds of mullet (*Mugil*), cat-fish (*Copidoglanis tandanus*), and in one northern river the Murray cod. Besides these there are occasionally both bream and flathead in freshwater. The bull-rout, gudgeons, &c., have been mentioned as far as their importance or interest required.

# CHAPTER XI.
## The Fish Market.

THE following particulars are taken from the Commissioners' report. The alterations which have taken place since that time will be referred to when dealing with the Fisheries Act, whose provisions can hardly yet be said to have come into full operation :—

The Fish Market of Sydney, to which all fish intended for sale are brought, is under the supervision of an Inspector of the Municipality, whose duties are to ascertain the fitness for human food of the fish brought into the Market, to condemn and cause the instant removal of that which is unfit, and to dispose of the remainder by public auction. The duties of this officer commence in the summer months at 5, and in the winter at 6 a.m. The disposal of the fish by public competition is an arrangement made by the Inspector and assented to by the fishermen, and appears to be the most desirable mode of dealing with the supply. The fish are brought into Market during the night and in the morning immediately before the sale. The supplies which are brought from a distance in coasting-steamers arrive generally during the night previous to the day's sale. A return, carefully prepared by Mr. Inspector Seymour, gives abundant information concerning the sources of supply, the varieties and quantities and value of the fish, extending over a period of seven years. During the winter months these supplies are enlarged by the facilities for bringing in a fit state to market the produce of the fisheries of places as distant as Port Stephens to the north and Jervis Bay to the south. As no arrangements have ever existed for the carriage along our coasts of fish in ice, any distant fishing-grounds are during the summer months proportionally valueless for the purposes of a supply to the metropolis. Up to a very recent period, no arrangements had been made for the reception in an ice-house at the Market of the fish as they arrived. The consequence of the entire absence of any means of preserving the fish after capture until their delivery at the Market, and there up to the time of their disposal, was the loss of very large quantities of valuable food. The particulars of this loss will be found in the return made by Inspector Seymour, to which reference has already been made. The result of the non-employment of ice or any other means of bringing the fish to Market in a fresh condition, was of course to limit the sources of supply to the harbour and to its immediate vicinity. It would appear that the Sydney Market is regularly supplied in the following way :—There are engaged in fishing in the numerous bays of Port Jackson, including the Parramatta River, twenty-seven seine-boats, each manned by four men ; and eight boats for line-fishing, with crews of about three each. Only one steam-vessel is regularly employed in the fishing trade, although in the winter months considerable quantities of fish are brought in the Hunter River and Illawarra lines of steamers from Tuggerah, Lake Macquarie, Newcastle, Port Stephens, and other places to the north, and from Wollongong, Shoalhaven, and Jervis Bay to the south. The solitary steam-vessel in the trade leaves Sydney every Monday morning at 7 o'clock, goes to Broken Bay, and returns to the Sydney Market at 5 the next morning, bringing daily freights up to Saturday morning. She collects her freights from fishermen at Broken Bay, and goes about 10 miles up the Hawkesbury River. It was ascertained that for this service her owners receive one-third of the proceeds of all fish carried by this vessel to market. The fishermen supplying the Market do not seem to have any direct dealings with the fishmongers or hawkers who are engaged in the distribution of the fish ; but the entire business between the producers and the fishermen is managed by agents, who collect the moneys from the buyers and pay them over with commission deducted, to the fishermen. As may at once be ascertained by a reference to the figures contained in the tables prepared by Mr. Inspector Seymour, the fish supply is quite unequal to the existing local demand, while with the immensely increased facilities for disposal created by the opening up of railway communication with places in the interior, where no supplies of fish can be obtained except from the seaboard, there seems no reason for supposing that under existing circumstances any such demand could be at all satisfactorily met ; and yet, as will be seen hereafter, the sources of supply, if properly guarded, are practically equal to any demand that could possibly be made upon them. At present the price of fish is, as will be seen from the evidence, excessively high.* In January of this year bream were fetching from 30s. to £2 a bushel (that is sold wholesale to the fish-hawkers) ; schnapper were readily fetching from 28s. to 30s. a dozen ; squires, from 5s. to 15s. a dozen ; whiting, from 6s. to 8s. a dozen ; garfish were sold at £2 18s. a bushel ; soles and flounders at times 1s. to 2s. per pair, and other varieties of fish at proportionate prices. The prices given by the consumers, who in the absence (with but a few exceptions) of regular fish-shops, purchase from the hawkers, are enormously increased—in some cases, as will be seen by the evidence, doubled, and in others quadrupled. The inadequacy of the supply to meet the demand may be inferred from the value of the imports as it is furnished by the Statistical Register. From this it would appear preserved fish of various kinds was imported to the value of £161,970 in 1877, and £133,334 in 1878. We ascertained that there are several places even in the city of Sydney itself where fish is rarely if ever seen, and where the people have become so entirely unaccustomed to the use of it as an article of food that they seldom if ever think of purchasing it.

* This was written in 1879.

## FISH AND FISHERIES.

The following return will show some of the increase in our fish supply in the past ten years. It will be noticed that the condemned fish becomes less on the whole:—

RETURN of Fish brought to the Eastern Market, Woolloomooloo, and quanity of fish condemned as being unfit for human food, for the years ending 30th November, 1873, 1874, 1875, 1876, 1877, 1878, and 1879.

| Net Fish, bushels. | Schnappers, doz. | King-fish, No. | Jew-fish, No. | Tuglers, doz. | Sea Mullet, doz. | Salmon, doz. | Rock Cod, doz. | Flathead, No. | Nannygey, doz. | Mackerel, doz. | Soles, doz. | Lobsters, doz. | Sweeps, doz. | Gropers, No. | Prawns, bushels. |
|---|---|---|---|---|---|---|---|---|---|---|---|---|---|---|---|
| 1873. | | | | | | | | | | | | | | | |
| 11,456 | 1,037 | 1,365 | 794 | 742 | 5,679 | 136 | 125 | 763 | 234 | 677 | 89 | 766 | 233 | 11 | 27 |

Quantity of fish condemned as being unfit for human food:—Net fish, 48 bushels; schnappers, 37 dozen; sea mullet, 594 dozen; prawns, 5 bushels.

| | | | | | | | | | | | | | | | |
|---|---|---|---|---|---|---|---|---|---|---|---|---|---|---|---|
| 1874. | | | | | | | | | | | | | | | |
| 13,427 | 1,736 | 951 | 1,197 | 637 | 984 | 232 | 24 | 600 | 27 | 188 | 132 | 150 | 33 | 15 | 59 |

Quantity of fish condemned as being unfit for human food:—Net fish, 36½ bushels; schnappers, 13 dozen; sea mullet, 43 dozen; prawns, 2 bushels.

| | | | | | | | | | | | | | | | |
|---|---|---|---|---|---|---|---|---|---|---|---|---|---|---|---|
| 1875. | | | | | | | | | | | | | | | |
| 12,899 | 2,804 | 607 | 1,012 | 825 | 4,250 | 134 | 40 | 2,943 | 219 | 10,823 | 401 | 196 | 49 | 9 | 118½ |

Quantity of fish condemned as being unfit for human food:—Net fish, 93 bushels; schnappers' 11 dozen.

| | | | | | | | | | | | | | | | |
|---|---|---|---|---|---|---|---|---|---|---|---|---|---|---|---|
| 1876. | | | | | | | | | | | | | | | |
| 12,922 | 2,443 | 748 | 1,116 | 763 | 4,367 | 214 | 13 | 1,188 | 78 | 10,506 | 170 | 247 | 51 | 12 | 119 |

Quantity of fish condemned as being unfit for human food:—

| | | | | | | | | | | | | | | | |
|---|---|---|---|---|---|---|---|---|---|---|---|---|---|---|---|
| 1877. | | | | | | | | | | | | | | | |
| 12,071 | 1,201 | 1,114 | 643 | 1,055 | 4,315 | 106 | 20 | 680 | 219 | 415 | 98 | 156 | 500 | 12 | 37 |

Quantity of fish condemned as being unfit for human food:—

| | | | | | | | | | | | | | | | |
|---|---|---|---|---|---|---|---|---|---|---|---|---|---|---|---|
| 1878. | | | | | | | | | | | | | | | |
| 12,907 | 1,721 | 1,032 | 1,324 | 583 | 1,379 | 374 | 19 | 1,123 | 31 | 217 | 102 | 379 | 42 | 19 | 114¾ |

Quantity of fish condemned as being unfit for human food:—Net fish, 110 bushels; schnappers, 3 dozen; sea mullet, 48 dozen.

| | | | | | | | | | | | | | | | |
|---|---|---|---|---|---|---|---|---|---|---|---|---|---|---|---|
| 1879. | | | | | | | | | | | | | | | |
| 12,642 | 1,689 | 685 | 1,423 | 516 | 1,412 | 311 | 17 | 1,250 | 27 | 28 | 113 | 453 | 51 | 15 | 11 |

Quantity of fish condemned as being unfit for human food:—Net fish, 57 bushels; schnappers, 13 dozen; sea mullet, 44 dozen.

GRAND TOTAL of fish brought to the Eastern Market, Woolloomooloo, with quantity condemned as being unfit for human food to 1879.

| | | | | | | | | | | | | | | | |
|---|---|---|---|---|---|---|---|---|---|---|---|---|---|---|---|
| 88,334 | 13,531 | 6,502 | 7,508 | 5,123 | 17,304 | 1,707 | 258 | 8,600 | 875 | 23,036 | 1,105 | 1,609 | 1,700 | 93 | 486 |

RETURN of Fish brought to the Eastern Market, Woolloomooloo, with amount realized at auction sales, market revenue, and quantity of fish condemned as being unfit for human food, for the year ending 30th November, 1880.

| Net-fish, Bushels. | Schnapper, doz. | King-fish, No. | Jew-fish, No. | Traglers, doz. | Sea Mullet, doz. | Salmon, doz. | Rock Cod, doz. | Flathead, doz. | Nannygey, doz. | Mackerel, doz. | Soles, doz. | Sweeps, doz. | Lobsters, doz. | Groyers, No. | Green Prawns, bushels. |
|---|---|---|---|---|---|---|---|---|---|---|---|---|---|---|---|
| 12,927 | 1,715 | 702 | 1,150 | 611 | 1,308 | 423 | 22 | 1,322 | 70 | 57 | 134 | 63 | 519 | 16 | 22 |

Amount realized at auction sales from December, 1879, to November, 1880, inclusive, £16,047.

Market revenue—Auction fees on sale of fish, £803 12 1
Rents, &c.. £149 10 0
———————
£953 2 1

Quantity of fish condemned as being unfit for human food :—
Net fish, 49 bushels ; schnapper, 2 dozen ; sea mullet, 39 dozen.

RETURN of Fish brought to the Eastern Market, Woolloomooloo, with amount realized at auction sales, market revenue, and quantity of fish condemned as being unfit for human food, for the year ending 30th November, 1881.

| Schnapper, dozen. | Net Fish, bushels. | Traglers, dozen. | Flathead, dozen. | Sea Mullet, dozen. | Salmon, dozen. | Nannygey, dozen. | Eels, dozen. | Soles, dozen. | Mackerel, dozen. | Sweeps, dozen. | Jew-fish, No. | King-fish, No. | Groyers, No. | Lobsters, dozen. | Crabs, dozen. | Live Prawns, bushels. |
|---|---|---|---|---|---|---|---|---|---|---|---|---|---|---|---|---|
| 1,734 | 15,194 | 316 | 300 | 2,811 | 523 | 68 | 50 | 121 | 22 | 46 | 1,491 | 564 | 38 | 443 | 78 | 134 |

Amount realized at auction sales, £17,949 14s. 7d.
Commission on sale of fish—£899 5s. 5d. ; rents, &c., £136 15s.—£1,036 0s. 5d.
Fish condemned :—Net fish, 258 bushels ; schnapper, 61 dozen ; lobsters, 25 dozen ; salt fish, 60 dozen.

## CHAPTER XII.
### *The Development of our Fisheries.*

THE following valuable and most suggestive observations are taken almost verbatim from the Report of the Royal Commission. It is not too much to say that nothing at all equal to them has been published in the Colonies. They comprise information from the most competent and most experienced persons on the subject of Fisheries which the Colony possesses. It is to be hoped that they will be studied, and made use of by those engaged in the fishing industry.

We proceed then to offer a few suggestions as to the best mode of utilizing and improving the vast resources which the sea offers us. Fisheries are of two classes—those which are undertaken for the supply of the daily demand, and those which are undertaken at fixed periods, and for a short time only. To the first of these we shall give the name of "Ordinary Fisheries," to the other "Special Fisheries." The "Ordinary Fisheries" may be divided into "deep sea or line fisheries" and "harbour or net fisheries."

Deep sea fishing is at present carried on in open boats, manned generally by three fishermen. These people have no means of preserving from putrefaction the fish they catch, so that there is no temptation to them to go out to any of the many fishing-grounds off the coast, even if the description of boat in use were fitted for that purpose. But the very reverse is the case. The boats are small, afford no comfort or protection from wind and weather to the crew, and in fact cannot be used with safety except on a fine day and in close proximity to the land. Suggestions have often been made for the employment of decked boats of considerable size, with a well for the reception of the fish—a kind of vessel much used in the Channel and North Sea fisheries of England; and we believe one or two such boats have actually been tried here, but entirely without success. Our fishes it would seem, the schnapper more particularly, will not survive in the well above a very short time, and are found to be bruised and made unfit for food by the process.

The general use of ice, and the cheapness of its production, render now unnecessary such expedients as well-boats. Strong roomy vessels of 100 to 150 tons burden, with moderate steam-power and a good ice-house, and manned by ten or twelve fishermen, are what is best suited to our climate and coast. It might be unnecessary to have these fishing-vessels so large, or it might be advantageous to increase their tonnage; these are matters of detail not necessary for our purpose, which is merely to point out how our best fishing-grounds can be reached with comfort and safety to the fishermen and secure the perfect preservation from putridity of the fish. The steam-power, which would be only used as an auxiliary in going out to and in from the fishing-grounds when the winds were light or unfavourable, should be equal to a speed of at least 6 knots an hour. The engines and all connected with the victualling and bedding of the crew should be in the stern of the vessel, and the forepart, both on deck and beneath, should be given up entirely to the curative processes for the fish caught and for the ice-house.

Fishing-vessels of that description could with safety go out to sea in any weather and take supplies and all sorts of fishing-gear for a week's consumption; the crew should be experienced fishermen, and if possible be themselves the proprietors. The cost of the vessel with a complete outfit of everything required would probably be little short of £4,000. The expenditure weekly, including £2 a week for the wages of each of the crew and interest at 10 per cent. on the capital invested, would be (say) wages £24, interest £8, add £10 for losses and wear and tear—£42 per week. This is taking the expenses at an extreme figure; a moderate calculation of the week's receipts would show a much larger return. In the course of the week it is pretty certain that twelve fishermen with good appliances will catch at least 4 tons of fish, and taking these at the low estimate of £20 per ton—less than half the price of salted and dried ling—the weekly profit would be nearly 100 per cent. on the outlay. Besides, we may fairly calculate that the best of the fish will, as they are caught, be at once cleaned and placed in the ice-house, and these of course will realize a much higher price than above estimated. With vessels too, so commodious as these, everything may be utilized and nothing should be lost; the inferior fish, or those not worthy of being put in ice, should be salted on the spot, the air-bladders should be washed and dried at once, particularly those of the jew-fish, teraglin, and silver eel, and even the sharks may be made a source of profit by the saving of the fins and the oil from the liver. The prime fish, such as black rock-cod, schnapper, nannygey, gurnard, Sergeant Baker, &c., brought in the ice-chests of these vessels would of course require to be received into a cool room the moment of their arrival; but the advantages attending the use of ice are becoming so fully recognized now, and the ice itself is to be got at such a moderate price, that there is little doubt that every fishmonger will be amply supplied with all that is necessary to keep the fish fresh. It should be understood that it is exceedingly injurious that fish should be ever frozen, but kept at a temperature of 32° F. they undergo no change, and retain all the excellence of the most freshly caught fish.

Some improvements on the present mode of line-fishing might also be attempted. On the Newfoundland banks, and in the North Sea fisheries of Norway, the "bultow" is much used for the capture of the cod and other fish. The following description of the "bultow," as used by the French fishermen at Newfoundland, is taken from "Simmonds' Commercial Products of the Sea," page 27:—"The bultow is a long line, with hooks fastened along its whole length, at regular distances, by shorter and smaller cords, called 'snoods,' which are 6 feet long, and are placed on the long line 12 feet apart, to prevent the hooks becoming entangled. Near the hooks these shorter lines, or snoods, are formed of separate threads, loosely fastened together, to guard against the teeth of the fish. Buoys, buoy-ropes, and anchors or grapnels are fixed to each end of the line, and the lines are always laid, or, as it is termed 'shot,' across the tide, for if the tide runs upon the end of the line the hooks will become entangled, and the fishing will be wholly lost—for the deep sea fisheries the bultow is of great length. The French fishing-vessels, after anchoring on the bank, in about 45 fathoms of water, run out about 100 fathoms of cable, and prepare to catch cod with two lines, each 3,000 fathoms in length. The snoods are arranged as previously described, and the hooks being baited, the lines are neatly coiled in half-bushel baskets, clear for running out. The baskets are placed in two strong-built lugsail boats, and at 3 o'clock in the afternoon both make sail together at right angles from the vessel on opposite sides. When the lines are run out straight, they are sunk to within 2 feet of the bottom. At daybreak next morning the boats proceed to trip the sinkers at the extremites of the lines, and while the crew of each boat are hauling in line and unhooking fish the men on board heave in the other end of the lines with a winch. In this way 400 of the larger bank cod are commonly taken in a night. The fish are cleaned and salted on board, and stowed in the hold in bulk; the livers to be boiled for oil are put in large casks, secured on deck." We may observe that the French method mentioned above of treating the cod livers is not to be recommended, the extremely disagreeable smell of cod-liver oil being due entirely to the putrefied condition of the livers. To have a perfectly sweet and pure oil, it should be obtained by the moderate application of heat when in an undecomposed state.

The "harbour and net" fisheries, like those of deeper water, cannot be made productive to a much greater extent than at present, without increased facilities for bringing the fish fresh to market, and there is no more effective way of accomplishing that end than by the use of ice; and indeed that is the only suggestion we can make for the improvement and development of these harbour fisheries. The boats in use among these net fishermen are good and well adapted for the purposes they serve; the seine nets they use are undoubtedly unexcelled in securing any fish there may be within their circumference, and the means of getting their fish to market seems to be the only trouble to which the rather lazy net fishermen are subject. Ice and rapid communication are the only effectual remedies for this; but these will be quickly availed of as soon as sufficient inducements offer in the shape of immediate money profits.

The trawl net, the chief instrument of capture on the English coast, has never come into general use here, and it is probable never may to any great extent, owing to the rocky character of our coast. If, however, a detailed survey of our fishing-grounds should prove the existence of some sandbanks, the trawl net would be found to be the most useful and almost only mode of getting at such fish as the flounder, sole, and John Dory.

The "special fisheries" are those which like the salmon, the herring, the pilchard, the anchovy, the tunny, the mackerel, and other fishes of similar habit, appear at certain periods for a short time only, and therefore if they are fished for at all must be fished for and utilized in a very different way from the every-day fishes. They differ from the other fisheries also, in so far as the fishes of the special class have a value chiefly arising from some particular mode of preparing them for the local market or exportation—the numbers in which they are generally taken far exceeding the possible demand for them in a fresh state.

The sea mullet (*Mugil grandis*) is, of all our fishes, the one that gives greatest inducements for a special fishery. In the months of April and May it makes its appearance in very large shoals on our coast, never going far from land, always proceeding in a northerly direction, and

T

showing a disposition to enter every inlet and harbour in its course. It is then in the finest condition, and full of roe. Its annual migration at that period is simply in search of suitable spawning grounds. We are convinced that with proper appliances, and under proper restrictions, the mullet fishery at this time, lasting probably six weeks, might be made of very great value to the Country. The quantity which could be consumed in a fresh state during the fishery would be in a very small proportion to the numbers captured, and it would therefore be necessary, in order to utilize the vast numbers of this fine fish then offering itself for our use, to hit upon some means of preserving the larger portion for future use in a marketable form. At present it is not unusual to salt and smoke it, but its very fatness and excellence make it a bad fish for this mode of treatment—it takes the salt too readily, and is apt to become rancid. The roe, however, salted and smoked, is equal to anything of the kind in the world, and in that state is always rapidly bought up in any quantity.

It is evident that with a fish of such richness and delicacy no plan can be so good for preserving its excellence of flavour as that generally adopted in the case of the salmon—a fish possessing many of the same qualities—boiling and hermetically sealing in tin cans. The form of the tin need not be exactly the same as we are accustomed to see salmon in—in fact there is considerable variety even in that, the Dutch practice being to tin salmon in long cans holding each two full-sized fish. The process is extremely simple.

At the Columbia River (where in 1876 the quantity canned was 428,730 cases, each containing 4 dozen 1-lb. tins, or about 23,000,000 lbs.), they cut the fish up with a number of curved knives—say seven or eight—with a lever attached. The fish are laid on a bench, and the lever being pulled down, the knives cut the fish into sections the size of the tin, whether 1, or 2, or 4 lbs. There are more 1 lb. tins put up than those of 2 lbs. or 4 lbs. Then the fish are put into the tin, and a small tea-spoonful of salt is added to each can, to take away any unnatural flavour. The cans are then put into a large boiler and heated with steam, and after the fish have boiled a certain time, say twenty minutes under the greatest heat they can get—the steam is 210 degrees, but then they get the heat up to 280 degrees by the addition of chemicals.

They put salt into the water to enable it to be brought to a greater heat than the boiling-point of pure water. The salt alone would not bring it up to 280 degrees, but they use chloride of calcium. Then after the fish are cooked a certain time the screen that the cans are on is raised out of the boiler, and a man goes round them with a little mallet, having a small spike on it, and he taps every one of the cans, by which means the gas that has accumulated in the cans escapes. After the fish have settled down he solders the small hole, lowers the screen with the cans again into the boiler, and the water is raised to the boiling point, so as to produce steam inside the cans. When they are taken out and cooled the steam condensing produces a vacuum in each can, and the cans are then passed into a place where they are left for one, two, or three days, to see that they are perfectly tight.

Sooner or later other methods may be discovered of treating our fine species of mullet. Whether or no, we have no doubt that its fishery is destined to be a most important industry. The mode of capturing this fish is a matter which requires consideration. At present the fishermen catch them after they enter the harbours, bays, inlets, &c., and follow them up to their spawning grounds, scattering them and effectually interfering with them in the discharge of these necessary functions. Such a wanton destructive mode of fishing should not be tolerated. The breeding of the fish need not be interfered with, for they can be got in any quantity at the mouths of the bays or in the open sea. For this mode of fishing, drift nets, such as are used for the herring and mackerel fisheries, but with larger meshes, might be found to answer. But the description of net chiefly used in the menhado fishery, on the coast of Maine, is we think still more likely to be suitable. It is thus described by Mr. Simmonds in the "Commercial Products of the Sea," page 222 :—" The seines are made of strong cotton twine and are 130 fathoms long (780 feet), and from 80 to 100 feet deep. At the eastern end of Long Island, where the fishing is in deep water, the depth is even greater. Along the bottom of the seines run lines so arranged that they can be drawn up like an old fashioned purse—whence the name 'purse seine.' The top of the seine is attached to buoys of cork or wood, and these, when the whole is thrown into the water, hold the upper edge at the surface, while the remainder hangs vertically beneath it. The seine is loaded into two boats, which also form a part of the outfit of the yacht, and are always with her when not engaged in taking fish. Thus furnished the yachts start on a cruise in search of the fish, which go in immense schools. When a school is met with it is necessary to drop the seine in front of them, otherwise no fish would be taken, as they would swim away in front before the seine could be closed round them. The boats get ahead of the school and pay out the seine as they separate. When the school is fairly in the seine the boats come together and completely surround the fish. At the point where the boats first started a heavy weight called a 'tom' is attached to the bottom of the seine, and to this weight, which rests upon the bottom, are fastened the lines which 'purse' up the bottom and prevent the fish from escaping below. When the bottom is drawn together the men haul the seine into the boats, and shake the fish down into the bunt, as the purse formed by the seine is called. The fish are taken out of the seine into the 'carry-ways' by means of dip nets." The "carry-ways" are additional vessels attached to each yacht for taking the fish ashore.

A system also of having a look-out kept along the coast to the south upon the movements of the shoals of mullet, so that information might be telegraphed to the fishing stations, as is done in the case of the tunny in the Mediterranean, would be most useful as a guide to the people engaged in this fishery.

The "maray" (*Clupea sagax*) is a very rich, oily, well tasted fish of the herring family, which passes north along our coast about midwinter in enormous shoals. The same fish has been seen to pass south along the east coast of New Zealand in the month of February, so that there is every reason to infer that it is a migratory fish in the truest sense. Some idea of the vast extent of the shoals of these fish may be formed from the following quotation. Professor M'Coy, of Melbourne, in describing its various appearances in Port Phillip, says:—"After remaining for a few weeks they disappeared, until the same time in 1866, when they arrived in such countless thousands that carts were filled with them by simply dipping them out of the sea with large baskets. Hundreds of tons were sent up the country to the inland markets, and through the city for several weeks they were sold for a few pence the bucketful, while the captains of the ships entering the bay reported having passed through closely packed shoals of them for miles."

This fish is salted and smoked in some parts of New Zealand, and the highest award of excellence was adjudged a few days ago at the International Exhibition in the Garden Palace to the Picton bloaters prepared from this very fish. No attempt has been made in this Colony at any time we believe to make any use of this fish, and yet there are none which would yield to a little enterprise a more certain return. Even for oil and manure it would pay well to fish for them on a large scale. The "menhado" (also a kind of herring) fishery, on the coast of Maine, which is conducted solely for the production of oil was valued in 1873 at £325,000, of which £200,000 was derived from the oil and £125,000 from manure. But the "maray" seems fitted for a higher destiny than oil or manure,—the size, quality, and delicacy of this fish point it out as a worthy substitute for the sardine. The process of preserving fish in this way is very simple, and is thus described by Mr. Simmonds, in the "Commercial Products of the Sea," in his chapter on the sardine fishery of the Mediterranean:—"Brought to land they are immediately offered for sale, as if staler by a few hours they become seriously deteriorated in value, no first-class manufacturer coming to buy such. They are sold by the thousand; the curer employs large numbers of women, who cut off the heads of the fish, wash and salt them. The fish are then dipped into boiling oil for a few minutes, arranged in various sized tin boxes filled up with the finest olive oil, soldered down and placed in boiling water for some time to test the boxes, those which leak being put aside. It does not always seem to be remembered that the longer the tin is kept unopened the more mellow do the fish become, and if properly prepared, age improves them, as it does good wine; but if they are too salt at first, age does not benefit them—they always remain tough." The value of the sardine trade to France is about £700,000 per annum.

The maray must be caught in the open sea, as it does not, except by accident, enter our harbours. The fishing will require therefore to be conducted with drift nets, such as are used for herrings in the North of Scotland, or better still, by the purse seine, described on previous page.

The approach of the shoals from the south should in this case, as with the mullet, be watched and signalled.

There are other herrings on the coast capable of being utilized to an equal or even greater extent than the "maray," but the processes of catching and curing would be the same.

The mackerel visits our shores at intervals in very large shoals, so does the "tailor" and a host of other fishes, all of which may ultimately become special and important fisheries, for salting or other purposes, but at present the development of the most obviously useful kinds is all that can well be attended to.

# FISH AND FISHERIES. 149

At the Fisheries Office, Garden Palace, there are now on view some of the fishing implements used in other parts of the world. They were obtained by the Commissioners of Fisheries, with the objects of testing the applicability of the modes of fishing practised in some parts of the Continents of Europe and America to the fisheries of this Colony, and of affording to our fishermen an opportunity of acquainting themselves with the necessary appliances. Amongst the articles referred to, and which can be inspected by persons interested, will be found the following:—A purse seine, such as is used in the menhaden fishery on the Maine coast, North America; a trawl net, with beam, complete, as used by the Grimsby fishermen; a drift net of the kind used in the herring fishery, Scotland; a bultow, or set line, used by the French fishermen at Newfoundland; a set line, used by the Norwegian cod-fishers in the North Sea; glass baits and floats, in use in the Norwegian fisheries; an otter trawl net, a trammel net. In addition to these are a set of boat fittings for working the menhaden purse seine; also, a pair of woollen nippers, not well known to any but American fishermen; they are used in hauling the bultow. Herr Von Behr, President of the German Fischerei Verein, said in regard to these nippers that, though insignificant in appearance and simple in contrivance, they deserved to rank amongst the most important of the American exhibits at Berlin. It is possible they may be utilized in the deep sea fishing on this coast.

## CHAPTER XIII.
### Acclimatization and Pisciculture.

THIS department of natural history has assumed the most important aspects of late years, and may be said at the same time to have become one of the most popular and fashionable of zoological recreations. According to Dr. Günther, artifical impregnation of fish ova was first practised by J. L. Jacobi, a native of Westphalia, in the years 1757 and 1763, who employed exactly the same method which is followed now. The idea of course was to favor the natural processes, and by preventing the waste of ova to restore the failing supplies of rivers, ponds, and streams. The process is as follows:—The fructifying vessel, which is tolerably large and with a flat bottom, is filled about one inch deep with pure water, carefully kept at the same temperature as that into which the fish resorts for spawning. A full-roed healthy female fish having been selected, is taken with as much gentleness as possible by two persons, who hold it in a manner not likely to injure the fish, by the head and tail respectively. The fish is held in an upright position, and after it has ceased struggling, the roe, if it be quite ripe, commences to run from the vent, and in order to obtain as much as possible the fish's sides and belly are gently rubbed downwards. If the fish be large it is never worth while to press out all the roe at one time. In small fishes only about half the roe is taken, and the fish is put back again for a day or two into the reservoir, and then the process is repeated. "If the eggs are quite ripe they all pour out into the water following each other in rapid succession, like shot from a shot-belt. If the eggs do not come out quite easily, give the tail a gentle shake to loosen the eggs that remain in the abdomen; but recollect, if you use force you will spoil the experiment. The eggs *must* run out quite freely."—Buckland.

After the roe is pressed out it is spread over the bottom of the basin. A male fish is then taken, and a few drops of the milt is pressed out into the water. Pressure is made on the abdomen of the male in the same way as the female. If the milt is ripe, it will instantly discolour the water and make it white. The roe and milt are then stirred up with the finger, and left for about five minutes. The milk-coloured water is then poured off, and fresh water gently added until the eggs appear quite clear again. The milt of one male will fertilize the ova of many females, and the fish can be returned to the sea or river none the worse for the operation. Then the impregnated ova are either placed into the breeding boxes or packed for exportation. The manner of packing is this:—The boxes are about 6 inches in height and width, and 2 feet long. On the bottom is a layer of damp moss, mixed with ice (if the temperature of the season or the country require it). On this is spread a damp linen cloth, over it a thin layer of ova is laid. Another wet linen cloth and more moss completes the packing, which has to be rather tightly fastened to prevent shaking and friction. In many places in Europe no roe is packed until two black spots appear in the ova (the eyes), because the fish bear transport better in that state. If a long transit be anticipated, the ice arrangements must be different, because melted ice water injures the roe.

# FISH AND FISHERIES. 151

The simplest hatching apparatus consists of a series of shallow boxes half-filled with sand, and placed at a gradually descending plane from first to last. A very small flow of water from a tap into the uppermost box, and from that to the others, will keep the water oxydized and prevent the development of septic germs. It is found however that this arrangement has many disadvantages, and leads to a great waste of roe. The uppermost boxes consume the oxygen from the water, which by degrees loses the necessary vitality which is required for hatching the roe in the lower boxes. The method adopted in Dr. Klenertz's Fish-hatching Apparatus is this :*—Each box is supplied with it own stream of water, which is as much mixed with air as possible. The water is brought to a large stone trough in which the fish (trout) are kept during the summer, until the roe and milt are ready in the autumn. The water runs from this stone trough through numerous separate pipes to two other large troughs, one of stone and the other of wood, the sides and bottoms of which are covered with cement; each of these troughs is about 12 feet long, 2 broad, and 2 deep. Small gratings in the sides serve as rests for the hatching boxes. These are made of sheet-iron well painted in oil colours to prevent rust. They are a foot square on the bottom and 5 inches high. They are pierced with two rows of fine holes, and on the top there is a grove which is likewise furnished with fine holes. After the box is filled with clean gravel to the height of 3 inches, and on this the roe is thinly spread. The water which supplies these boxes is conveyed through small perpendicular tubes each furnished at the end with a rose so as to separate the falling water and bring it as much into contact with the air as possible. This water falls over the hatching boxes, and is still further aerated by the fine holes in the tops. The exclusion of much light is essential to the success of the experiment, especially of the hatching. The gravel is boiled for some time before using, to free it from the spores of plants, which are most destructive to the roe. This method has led to the best results, and only a small percentage of the ova are lost.†

Various modifications of these processes are in operation throughout Europe. Until it was adopted in Britain the salmon-fishery was in danger of extinction. The fecundity of fish is well known, but the waste was not. For instance, we find according to the late Mr. Frank Buckland, that the amount of ova to each pound of roe was as follows for the different fishes mentioned :—Salmon, 1,000 ova to each pound the fish weighs ; trout of 1 pound weight, 1,008 ; herring of half-a-pound 19,840 ; perch of half-a-pound, 20,592 ; jack of 4½ lbs., 42,840 ; mackerel of 1 pound, 86,220 ; sole of 1 pound, 134,466 ; brill of 4 lbs., 239,770 ; turbot of 8 lbs., 385,200 ; roach of three-quaters-of-a-pound-480,000 ; cod of 15 lbs., 4,872,000 ! Yet two-thirds of this or a larger quantity is lost under the most favourable circumstances. In some respects there is matter for congratulation in this, for it would not do to have the seas overrun with cod-fish, however

---

* This plan is not selected as the best, but as a convenient instance out of many which requires neither expensive buildings, nor apparatus.

† For a full description of the whole subject see "Fish-hatching," by Frank Buckland, London, Tinsley Brothers, 1863. A list of other works is given at the end of this chapter.

excellent the fish may be. But in those species in which the roe is large, such as the salmon, and the proportionate increase slow, the effect of fisheries rapidly tells. Thus it has been computed, according to given data and accurate calculations made by Messrs. Ashworth and Buist from the returns of fisheries, that only *one salmon's egg out of every thousand* deposited by the parent fish ever becomes a fish fit for human food. Other fish no doubt both in fresh and salt water suffer in proportion, the principal enemies being :—1. Floods and accidents to which the ova themselves are subject. 2. Fish devouring the ova, which includes in many cases the parent-fish. 3. Prawns and small crustacea. 4. Birds in fresh waters, ducks and swans especially. 5. Human enemies who kill or disturb the female fishes in spawning-time.

From nearly all these evils are the ova preserved by pisciculture, and it has now come to pass that there is not a single salmon river in Europe which does not have annually put into it a much larger number of artificially reared fish than is taken out of it. Up to a very recent date the most eminent establishment for pisciculture was at Huningue, near Basle. In this magnificent institution the eggs of fish are kept and advanced in their hatching until they arrive at the period at which they will bear travel. By this means many rivers in France have been actually re-stocked most abundantly with fish, employment given to hundreds of poor fishermen, and the food of the people greatly increased. The fish cultivated are the common trout, salmon-trout, lake-trout, Rhine salmon, Danube salmon, charr, grayling, and fera.*

The French Government has for the last thirty-five years turned its attention especially to Pisciculture, and may be said to have taken the lead of all other nations on the subject. In the commencement of the work the rivers and lakes of France were almost destitute of fish, but through the establishment at Huningue, they have been abundantly re-stocked; even though every man is allowed to fish with a line in all rivers and lakes, not on private property. The Huningue establishment was opened in 1852. The object in view was to stock the rivers with fish by the introduction of ova and young fry of the best kinds, and those of rapid growth. Over 20,000,000 of ova are partially hatched and sent away each year. " Working hard and enthusiastically," Mr. Frank Buckland says, " in the cause of improvement of fisheries are several French scientific gentlemen, to whom the highest possible praise should be accorded." Foremost amongst these is M. Coste, justly called the Father of European Pisciculture. It is owing to his exertions that millions of fish ova were sent not only throughout France but also to every European nation that requested them. Amongst these the British nation owes a debt to France for the aid thus afforded.

* A small salmon (*Coregonus acronius*), which when fully grown is 12 inches in length. Each fish produces a large quantity of eggs (ten to twenty thousand), but as it is very small the quantity is uncertain. Fera are called the herrings of the lakes, and are principally caught in the night. The young fish on leaving the egg can scarcely be seen in the water and passes through very small openings. In this way it escapes from hatching boxes so that its number cannot be correctly ascertained, but great quantities have been sent to stock the rivers and lakes of France. It lives in deep fresh waters and only frequents the borders of lakes to deposit the spawn. See Report of M. Coumes on Fisheries of France, 1861.

About the time that M. Coste was succeeding in his experiments Mr. Thomas Ashworth, of Cheshire, who with his brother was the owner of the Galway Salmon Fishery, commenced his operations. In a very short time—less than ten years—he had stocked with salmon various streams that previously had no salmon in them, as well as a district in their fishery 30 miles long by 10 wide.*

Scotland has also of late years done much for her fisheries. The establishment of Stormontfield, on the Tay, is now according to the late Mr. Buckland, a household word, and the observations both practical and scientific made by Messrs. Buist & Brown are of the highest importance. At a very small expense hundreds of millions of ova have been hatched and distributed. To give an example of what this has effected: in 1828 the rental of the fishery proprietors of the Tay was £14,574; it gradually fell off every year afterwards until 1852, when it reached £7,953, or nearly one-half; in 1853 the artificial rearing commenced; in 1858 the rental rose to £11,487, and in 1862 it had reached the value of 1828. This rise was not due to the increase of the value of salmon, because the increased price arose from scarcity, and the other fisheries which had not been re-stocked presented a gradually falling rental.† Things have improved generally throughout England, Scotland, and Ireland since those days. There salmon was rare and dear, and far in arrear of the demand, and though the London market alone has increased amazingly meanwhile, the salmon supply has kept pace with it. Not only that, but the rivers of Scotland especially are apparently more abundantly stocked than ever in this century, and rivers which had been cleared of salmon are now full again.‡ The last salmon caught in the Thames, says Mr. Buckland (writing in 1863), was nearly fifty years ago. He was caught at Windsor, and weighed 20 lbs. George IV bought him for twenty guineas. The mud and sewerage of the Thames has somewhat impeded the re-stocking of the river.

The salmon which we use so abundantly in the Colonies, and which comes to us in the well-known tin cases, is *Onchorrhyncus quinnat*, a fish which only differs from the salmon in the increased number of anal rays, which always number more than fourteen. All the species are migratory, ascending rivers flowing into the Pacific from the northern portions of the American and Asiatic continent. There are annually many millions of these fishes preserved, or as they call it "tinned," on the Sacramento and Columbia Rivers, but the supply is kept up by the artificial hatching and liberation of what is estimated to be two and a half millions of

---

* For a full account of these successes see "A Treatise on the Propagation of Salmon and other Fish," by Edward and Thomas Ashworth, London. Simpkin & Marshall, 1853.

† See Natural History of the Salmon as ascertained by the recent experiments on the artificial spawning and hatching of the ova and rearing of the fry, at Stormontfield, on the Tay. Arthur Hall, Virtue & Co., London, 1858. By W. Brown.

‡ The value of the salmon fisheries in 1871 was as follows:—England, £90,000; Ireland, £400,000; Scotland, £200,000. The sales at Billingsgate market for the same year were 1,764 tons of salmon, valued at £246,925, which is a little less than the average annual sale in London.

U

ova—at least so far as the Sacramento River is concerned.* In Germany the same efforts are made as in France, and in Sweden also the matter has been taken up very enthusiastically. It is in contemplation to commence something of the same kind in New South Wales, not only for the introduction of useful European fishes, but also for preservation of our own. At the present time the golden perch (*Ctenolates auratus*) has been much spread through the zeal and industry of Mr. Warren, of Wagga, who in the summer months transmits small boxes of ova to his correspondents.

It is necessary to mention here that fish ova are subject to many parasites which effectually destroy their vitality. These parasites are of different familes of fungi and algæ, especially a green conferva belonging to the genus *Œdogonium*. The ova are also very liable to the attacked by the small fungus which is so destructive to the common house-fly and affixes it to our window-panes on a little cloud of floculent threads. This is *Saprolegnia ferax*, and it is the one which seems to be destructive to many animals besides fish and flies. It stands on a debatable ground between algæ and fungi, a quality which it shares with three other genera which infest ova, and these are *Achlya*, *Pythium*, and *Aphanomyces*. The only remedy that can be suggested for this mould is to have the water used in hatching previously boiled and the gravel subjected to the same process, not in any case so difficult as at first sight it might appear. Weak solutions of Hyposulphite of soda have been suggested, but no results have been published. It is to be hoped that the Exhibition in 1883 will help to throw light on this question.

The use of pisciculture need not be confined to fresh-water fishes; there are many of our sea-fishes to which the art might be most profitably applied—the sea-mullet for one. What could be easier or less costly than to supply the drain and destruction of that most valuable fish by artificial breeding? Scoop out a shallow space of ground near the sea, the bottom about the low-water level, cut a channel of communication with the sea, so that every tide shall flow into it, put a piece of perforated zinc or wire netting across the channel so that nothing can get in to destroy the spawn; then get a few of the fishes on the point of spawning, squeeze out the roe gently with the hand into a tub with a little water in it, upon the roe so deposited squeeze the milt of the male fish, then stir gently with the hand so that every grain may come into contact with the male secretion, place the contents of the tub on the sandy bottom of the pond, and cover over with a net to prevent cranes and other water-fowl from disturbing the spawn. In the early spring the young fish will make their appearance, and after a few days' growth they may be turned out and left to their own resources. By the adoption of such a system the most extravagant waste of the old fish can be readily met. The full-sized roe of a single female are estimated to contain nearly 3,000,000 ova, so that the careful preservation of the yield of a very few fish would supply all the markets of the world. Under natural conditions the loss of spawn is enormous; by the artificial system it is reduced to a minimum.

There would be little difficulty in keeping up the numbers in the same way of any fishes for which the demand was so great as to risk the extinction of the fishery, provided they were fishes which spawned on our shallow beaches like the mullet. In that class would be included our best net fishes—the whitings, the garfishes, the sole, &c.—R.R.C.

It is as well here to draw attention to the larger proportion of eggs in the flat-fishes, which render them particularly favourable for pisciculture.

* See Report of the Commissioners for Californian Fisheries for 1878-79. The fish is generally believed to be English salmon. Also, Report of the Commissioners of Fisheries of Maryland from 1876 to 1880, where very full diagrams, plans, and drawings of the fish-hatching apparatus and houses are given.

# FISH AND FISHERIES. 155

The acclimatization of fish has not been carried out to a large extent in New South Wales. The river trout (*Salmo fario*) has been introduced with undoubted success into the rivers of Tasmania and Victoria, and into some of those of New South Wales. Shipments of salmon ova were made to Australia and Tasmania in 1849, 1852, 1860 and 1862, but none reached the colonies alive. After many experiments by Mr. Buckland as to the vitality of salmon ova when frozen for over 100 days, a shipment was sent away. In April, 1864, the first salmon ova (*Salmo salar*) arrived in Tasmania, and with them a few ova of the common trout, no salmon-trout being included in that shipment. In the spring of 1865 a number of smolts, estimated at about 1,500, went to sea. They had been very successfully hatched, and the loss was not great considering the enormous distance the ova had been brought. The packing was under the able superintendence of the late Mr. F. Buckland, and Mr. J. A. Youl, an old colonist residing in London,* and the whole success of the experiment was mainly due to the late Mr. Morton Allport, whose untimely death (September 10th, 1878) has been an incalculable loss to every department of science in Tasmania. In the same year that the smolts went to sea in Tasmania (1865) about thirty common trout were liberated in the river Plenty, and about 150 were retained in the breeding pond. In the following spring of 1866 the remainder of the salmon smolts from the first shipment, about 1,000 in number, took their departure for the ocean. Several grilse of 5 lbs. weight were reported subsequently in the fresh waters of the river Derwent.

In May, 1866, a second shipment of salmon ova arrived, and with them a number of salmon-trout ova. From this shipment, as far as regards the salmon, the hatching was very successful, and some 6,000 smolts were liberated in 1867 and 1868. The salmon-trout ova were not so successful. Only 496 were hatched, and of these more than 100 died before they reached the smolt stage. Of the survivors, 100 were permanently retained in a breeding pond, so that something less than 300 salmon-trout were liberated, while 8,500 salmon were sent forth. In October, 1869, two salmonoids were caught, one of which Dr. Günther pronounced to be a salmon-trout. In December, a third was caught which was more developed, having been six weeks longer in salt water. This was pronounced by Dr. Günther to be a true salmon. Since 1869 the number and size of the salmon captured have steadily increased. The late Governor, Sir Frederick Weld, frequently captured specimens of over 10 lbs. weight. It is still maintained by a few sceptical individuals that the fish caught are salmon-trout, and not salmon, but as the habits of the two migratory species, (*Salmo salar* and *Salmo trutta*) are similar, as they inhabit the same rivers and coasts, one species was just as likely to succeed as another in Tasmania. However, as about 8,500 salmon have been liberated to about 300 salmon-trout, and of the salmon 2,500 have had two years' start of the salmon-trout, it is manifest that the large fish captured are more likely to be true salmon, and that the experiment as far as Tasmania is concerned is a complete success. Already they afford excellent fishing, and at the falls and weirs on the Derwent the number of the fish making

---

* The boxes of ova were buried in ice sent for commercial purposes. The ova were in boxes as before described.

the leap is yearly increased. In the *Hobarton Mercury* of July 6, 1878, we read an account from Messrs. Allport and Read, of the capture of a splendid female salmon 20 lbs. in weight, 2 feet 11 inches long by 20½ inches in girth. It was taken in a spawning bed in the River Plenty, with a view of obtaining ova for hatching; but she had parted with most of the spawn, and only about 1,000 eggs could be obtained. The male fish on the same rid weighed 14 lbs.

The "Crucian carp" (*Carassius vulgaris*) is now abundant everywhere about Sydney. The gold-fish (*Carassius auratus*) is also completely acclimatized; and in Tasmania they have in addition the perch of European rivers (*Perca fluviatilis*) and the gudgeon (*Gobio fluviatilis*). These are all the fishes which have been introduced from other countries; but a little has also been done in the transfer of the fishes of the fresh waters of one part of the country to another. One species of the Murray cod was many years ago introduced into Lake George by the late Sir Terence Aubrey Murray, with the most astonishing success; the lake and all the creeks running into it are now fully stocked with cod of large size and of excessive fatness—the "marami," which abounds there forming their principal food. The Wollondilly River and Mulwarree Ponds near Goulburn have more recently been supplied with the young cod from Lake George, and it is said that the increase of the fish in these rivers has been most rapid and satisfactory. There can be no doubt that the transfer from one part of the country to another of the best of our fresh-water fishes is a much more sensible and feasible proceeding than the introduction with much trouble and at great expense of some of the most useless fishes of the European rivers. Old associations, however natural, connected with the names of gudgeon, carp, perch, &c., should not induce us to stock our rivers or fish-ponds with such inferior fishes. But it is quite possible to go too far even in the better direction of the transfer of our western fishes to our eastern waters. The Murray cod, as before observed, is a most destructive fish, swallowing up everything that comes in its way, not excepting its own species; and its introduction into new waters might result in the final destruction of other kinds. Some caution, therefore, should be used in the introduction of such a formidable fish; more particularly when from the same near source we can get supplied with fish of much better quality and less destructive tendencies. We allude to the fishes known in the Murrumbidgee as the "golden perch" and "silver bream"—the *Kaakaaluin* and *Koubery* of the aborigines—the *Ctenolates auratus* and *Therapon richardsoni* of the ichthyologist. For the purposes of food and all other objects to be attained by the stocking with good fish of our eastern rivers and fish-ponds, these fishes are infinitely superior to the cod, and their ova can be obtained for transport with ease.—R.R.C.

The colonists of Victoria and New South Wales have been over and over again recommended to introduce the Gourami (*Osphromenus olfax*), belonging to a family of fishes (Labarynthici) which have a kind of supernumerary gill in a cavity to itself which enables the fish to live for some time out of the water. In the accessory branchial cavity there is lodged a laminated organ which evidently has the function of assisting in the oxygenization of the blood. This accounts for the ease with which some can be acclimatized and transferred from pond to pond. The climbing perch which ascends trees belongs to this family. They are all cyprinoid fresh-water fishes belonging to the equatorial zone. The Gourami is reported to be one of the best flavoured fresh-water fishes in the East Indian Archipelago. Its original home is Java, Sumatra, Borneo, and several other islands, but it has been transported to Penang, the Mauritius, and even South America, and successfully acclimatized. It is very tenacious of life, and will eat anything and soon becomes very tame. It attains the size of a large turbot. A second, but much smaller species, *O. trichopterus*, is kept like the goldfish, for ornament. Its colours are of every hue, metallic in lustre, and of great beauty, but it is a very pugnacious fish.

Among the fishes which we would most strongly recommend to those who are specially interested in the subject of the acclimatization of useful fishes, we would particularly mention the salmon of the western coast of North America (*Oncorhyneus quinnat*), of which the chief fisheries are in the Sacramento and Columbia Rivers. The experiment of their introduction has been tried in New Zealand by Dr. Hector, on behalf of the Government of that country, we believe with success. It is the only salmon known which we in New South Wales can possibly, with our climate, have any chance of acclimatizing. We believe it is likely to do well in our Australian rivers, and in quality it does not seem, from the tinned specimens so largely imported here from California, to be one whit inferior to the salmon of the north of Europe. One other fish we would suggest to those who are disposed to introduce and acclimatize fishes of very superior quality. The *Chanos salmoneus* is of the herring family, of large size (2 feet long), extreme beauty and metallic brilliancy, and of the most exquisite flavour. It is found, though rarely, in these latitudes, its true habitat being in warmer seas. This fish is cultivated and kept in tanks in Southern India and Malacca, where it is highly prized, and regarded as a most expensive luxury. It might be tried in some of our northern rivers, the most evident objection to the experiment being the fact that the fish must have occasional access to salt water, and that once out of his river he might never return; but after all, the same objection might be urged against salmon culture. —R.R.C.

It should be mentioned that this fish has, like the Gourami, an accessory branchial organ which renders it especially adapted for acclimatization. Dr. Günther says that it attains the length of 4 feet.

In conclusion, a few words on artificial ponds and stocking them. The pond which is made by a dam should cover 10 or 12 acres at the least. The sides should be steep, except on one side, the depth over 7 or 8 feet, not only to allow for evaporation but also to prevent its becoming clogged with a growth of weeds from the bottom. There should be waste weirs at each end of the embankment for flood-times. The pond may be divided off by an embankment below the surface so as to portion off a shallow breeding-place. Trout will require a gravelly bottom, and will not thrive without it.

Carp (*Carrassius*) are the best fish for a pond. They breed often and abundantly, and their young are hardy and grow rapidly. The male is mature in five years, but the female not until she attains the age of eight years.

Tench (*Tinca tinca*) is another useful fish which will agree perfectly with carp. They will not attack or devour each other's spawn. The female is wonderfully prolific, over a $\frac{1}{4}$ of a million ova have been counted in one roe. Tench are said to hybernate in winter; its ova are of a green colour.

Perch (*Perca fluviatilis*) are ravenous fishes which will often devour their own ova and always that of other fishes, and therefore it is not desirable to keep them except in ponds by themselves. It is a poor table fish, but affords good angling.

Trout (*Salmo fario*) are also very ravenous, and often destructive to its own spawn.

Pike (*Esox lucius*) can never be safely introdoced near other fishes, as it devours all the small and weak ones, and will even attack young ducks. They are excellent for the table, but must be kept by themselves.

Roach (*Leuciscus rutilus*) is a pretty fish and feeds fast, being very much like carp in its habits and temper, but it is more use for ornament than for the table, and therefore is never spoken of as a fish for acclimatization. The above fish are common, hardy, inexpensive, and easily obtained. A full list on the subject would be a bulky affair. For the rest the works of Francis on Fish Culture may be consulted.

It should be mentioned that there are two especial enemies of the ova of fish which must not be overlooked. One is the stickleback (*Gasterosteus aculeatus* and *spinulosus*) and the water-beetle, *Hydaticus*, of which we have five or six species in our ponds. The stickleback is not known to be acclimatized, but might easily be accidentally introduced. No fish cultivation can proceed in the presence of two such enemies.

Fish may require feeding, and if so, may be assisted by throwing earth, worms, steeped grain or ground malt, and offals of poultry. Larvæ of the common blow-fly can easily be bred and thrown to them, as also house flies. With a long-pointed net large quantities of these can easily be collected, scalded, and thrown to the fish.

Besides fresh-water, salt-water ponds may also be constructed. They must of course be in places where sea-water can enter at half-tide. The sea-water is to be introduced by means of a sluice at this depth with a proper grating. By this means a regulated amount of water can be maintained at low-tide, and 8 or 10 feet at flood, and the water will be always changing. A great many of these ponds have been constructed in Europe. The fish kept in them were those of the neighbouring seas, such as turbot, sole, brill, plaice, flounder, rays, skate, herring, salmon, salmon-trout. The depth was usually not much over 12 feet or less than 3 at low water. Various other fishes have been tried, but the above enumerated have succeeded the best. Turtles have been kept in such ponds. The food generally consisted of butchers' offal mingled with blood, besides periwinkles, shrimps, and prawns. The fish in general do not improve by being kept in the ponds, they often become blind for want of shelter. Cod-fish and flat-fish thrive well for a time, as well as the rays and skate. Haddock also does well. There is no reason why the experiment should not be tried with our flat-fish. Those which we catch are all small, and might be reared to a considerable size.

The ancients were much more advanced in fish culture than ourselves. Artificial pieces of water for fish-ponds are of great antiquity. From the Egyptian paintings we see that those people used them, and they were in common use amongst the Greeks and Romans, coming down even to mediæval times in connection with monastic institutions. M. T. Varro, in his book "De Re Rustica," and Columella, in a work with the same title, both enter fully into the methods of constructing and preserving fish-ponds and salt-water vivaria. Columella's book is exceedingly interesting and enters into the fullest particulars, but as it may not be within reach of most readers they can consult the third chapter of the Rev. Dr. Badham's Prose Halieutics, or Ancient and Modern Fish Tattle (London, J. W. Parker, 1854). In this they will find ample details, even to the feeding of the stock and the dietary scale. The most extraordinary announcement in Columella is that the Romans turned lakes and rivers into natural vivaria; they placed fish and fish-spawn in them, even the ova of marine fishes which were known to ascend fresh-water streams. Thus they stocked, with perfect success, several large rivers which he names.*

The following list includes the most important part of the literature of Pisciculture:—

Fish-hatching; by Frank Buckland. London, Tinsley, 1863.
Instructions Pratiques sur Pisciculture; par M. Coste. Paris, Masson.
Pisciculture Pratique; par M. G. Millet. Bordeaux, Gounouilhou.
Pisciculture Pratique; par M. G. Millet. Paris, Siège de la Société, Rue de Lille, 19.
Pisciculture: Rapport sur le repeuplement des cours d'eau, &c.; M. Millet. Paris, Goin, Quai des grands Augustins, 41.
Pisciculture: Considérations générales et pratiques sur la Pisciculture Marine; M. G. Millet. Paris, Gras et Donnaud, Rue des Noyers, 74.
Notice Historique sur l'Etablissement de Huningue. Strasbourg, 1862.
Rapport sur la Pisciculture et les Pêches Fluviales en Angleterre, &c.; par M. Coume. Strasbourg, 1862.
Treatise on the Propagation of Salmon; by E. and T. Ashworth. London, Simpkin & Marshall.
The Natural History of the Salmon, as ascertained by recent experiments at Stormontfield; by W. Brown. London, Arthur Hall, Virtue & Co.
Fish Culture; by Francis Francis. London, Routledge. Supplementary Report on the Rivers of Spain and Portugal. Manchester, Love, & Barton.
The Salmon and its Artificial Propagation; by R. Ramsbottom. London, Simpkin & Marshall.
Translation of the Proceedings of the French Pisciculturalists. New York, Appleton.
Artificial Spawning, Breeding, &c.; by G. Boccius. London, Van Voorst.
Report of the Commissioners of Fisheries in Maryland, U.S. Annapolis, 1876 to 1880.
Harvest of the Sea; by J. G. Bertram. London, Murray, 1873.
Report of the Commissioners for Californian Fisheries for 1878-79. State Printing Office.
Art. Pisciculture in Encyclop. Brit.

* Most excellent amusement and instruction can be promised from a perusal of Badham's book, where the subject is dealt with in the most learned way, yet one sparkling with wit.

## CHAPTER XIV.
### Fishery Laws and Regulations.

A BILL "to provide for the development and regulation of the Fisheries of the Colony" as introduced by Sir Henry Parkes on the 13th January, 1881, in the Legislative Assembly, and of which the present Act is the outcome, was founded, in its most important principles, upon the Report of the Royal Commission on the Fisheries, dated the 3rd May, 1880 ; and, in respect of the Oyster Fisheries, the Bill adopted such recommendations of the Oyster Culture Commission contained in their report dated the 3rd May, 1877, as appeared to fall within the scope of the new system of administration which it was proposed to establish.

Some details of the Bill as introduced were modified in Committee, both in the Assembly and in the Council, and it was found advisable, in Committee, to incorporate an entirely new series of special clauses dealing with proprietary or private fisheries. These now form Part III of the Act as passed.

To understand the system of regulations introduced by this Act the short and admirable compendium of Mr. Alexander Oliver (one of the Commissioners of Fisheries), is placed before the reader. He says* :—
The sea-fisheries on the coasts and in the seas of Northern Europe (including in that expression Great Britain) are concerned chiefly with the capture of cod, herrings, mackerel, whiting, ling, haddock, and the flat fish, such as turbot, brill, soles, &c. the American, or rather the United States sea-fisheries are for cod, mackerel, menhaden (a species of herring), halibut, and haddock. Now, with the exception of the herring, which is known to spawn in firths and other inlets of the sea (for whitebait caught at the mouth of the Thames is now known to be the fry of the ordinary herring), all the other marine fishes here mentioned either spawn on banks away from the foreshores, and sometimes hundreds of miles distant from land, or else in places almost wholly inaccessible to drift-nets, seines, or trawls. Nets and trawls are therefore almost wholly incapable of destroying the young fry in the European seas ; and on the American coast the purse-seines, by which the mackerel and menhaden are caught, are quite innocent of any similar mischief. Of course the baited lines by which cod, haddock, whiting, and halibut are caught, whether in European or American seas, can do no harm either to spawn or youg fry. Why, then, should there be any statutory protection of these fish from legitimate capture, even if such protection were practicable ?

In this Colony, however, the conditions are quite different. The most valuable of our fishes are those which are most abundant and most easily captured. Of these the sea mullet, schnapper, the breams, the gar-fish, the black-fish, and whiting may be mentioned as the most important in an economic point of view ; and all these fish either always spawn in quiet waters at the head of salt-water inlets and in lagoons, or

---

* The Fisheries Act of 1881, &c., with an Introduction, Summary, and Index, by Alexander Oliver. Sydney, 1881, Government Printing Office.

# FISH AND FISHERIES. 161

do so when they can.* The bays and creeks of every harbour and the still waters of every lake and river swarm, under natural conditions, with the young of these fish in all stages of development. The fishing-net in use among our fishermen—a seine of almost unlimited length, and having in the centre a mesh almost small enough to catch shrimps—has long been a most destructive implement to these young fish ; and the constant harassing of the flats and beaches by "stalling" and the ordinary hauling net, has resulted not only in driving away the full-roed fish from their favourite haunts on the shallows into deep water, where the spawn is often shed under compulsion, and of course is rarely hatched, but also in something like extermination of the young of all the best kinds of our net-fish.

Our largest supplies of fish come from inlets, and not from open sea fishing as in Europe and America. In these truly antipodean divergencies from the conditions obtaining in other countries lies the justification of our recent legislation. We must protect the spawning fish and their young fry in the inlets, if we desire to prevent the absolute extinction of the best of our food fishes.

The Fisheries Act, 1881, is, without doubt, the first formal attempt at comprehensive legislation based on the principle of protecting the natural supplies of fish (including oysters, lobsters, and prawns under that term), and of regulating and controlling their capture. The administrative authority, subject to the customary executive control, under which the new system is to be worked, is a body of five Commissioners appointed by the Governor in Council, whose term of office is five years, and whose jurisdiction extends over the entire territory. The fisheries on the seaboard are distributed into three divisions—the Home, the Northern, and the Southern fisheries ; in one or other of which divisions every marine fishery, whether for fish, oysters, lobsters, or prawns, will be contained. The regular supervision of these fisheries will be the duty of inspectors and assistant inspectors ; and, in addition to the regular staff, certain Government officials are, by the Act, created inspectors *ex officio*. The inspectors and assistant inspectors are required to report to the Commissioners, in detail, at least once in every month, or oftener if directed, as to the state of the fisheries included within their respective divisions. The most extensive powers of framing regulations on all matters of detail are vested in the Governor in Council. The character and importance of this feature in the new system have already been adverted to.

* It is not perhaps correct to say that the schnapper spawns in inlets as a rule, for there is evidence of the spawn of this fish having been found on the school grounds at sea. It is, however, certain that the young fry of the schnapper resort to the inlets at a very early stage of development ; indeed, every inlet on our coast seems to be the natural asylum of the young of all our best food fishes.

The chief protective, regulative, and penal provisions of the Act are shown in the subjoined summary :—

## SUMMARY OF THE FISHERIES ACT, 1881.*

### Fishing Nets.

In all *Tidal-Water* Fisheries, that is to say, all sea-fisheries on the coast, unless during the *close months* or where the fisheries are absolutely closed against fishing-nets of every kind, a *lawful net* may be of a length sufficient to enclose a space of 300 *yards* measured along the corks, but must not have a *mesh*† in the *bunt* less than 2¼ *inches* or 3 *inches* in each *wing*. But *bona fide garfish nets* (if used during the *open* months, or elsewhere than in a fishery absolutely closed under section 17) will not be illegal if each wing is equal in length to the bunt, and if the mesh in each wing is not less than 2 *inches*, the bunt and wings being hung on the same cork and lead lines; but such garfish nets must not in the bunt exceed 30 *fathoms* in length, or have in the bunt a *mesh* less than 1½ *inch*. *Prawn nets* will be legal, under the same limitations as to time and place, if they do not exceed 15 *fathoms* in length, or do not have a mesh less than 1 *inch*.

The new Act contains a saving clause in favour of nets which under the old Acts would be legal during the "winter months"; that is to say, nets of 1¼ *inch mesh* in the *bunt* and of 2 *inches* mesh in the wings—the *bunt* being limited to 30 *fathoms* and each *wing* to the same length—may be used up to the 31st July, 1882, for the purpose of catching *garfish* only ; but such nets must not be used in *Close Fisheries* during the *close months*, nor used *at all* in fisheries absolutely closed against net fishing of every kind under section 17.

*Stalling* is prohibited at all times and places, but *bona fide meshing nets* not exceeding 60 *fathoms* in length, and not less than 4 *inches* in the *mesh*, are permitted if not set during the *close months* in any *close fishery*, or in any fishery absolutely closed under section 17.

Drift nets and purse seines of any length or mesh may be used in the open sea, *i.e.*, outside the mouths or limits of any bay, river, or inlet.

For *Inland Waters* a net may be *any length*, but must not be less in the *mesh* than 3 *inches*, and must *not to be set wholly across a river or creek*, but this restriction does not apply to nets used in private fisheries.

### Close Season for Net Fish.

The Act provides a close season extending from the *first of April* to the *thirtieth of September* in each year, during which no nets must be used in any tidal waters proclaimed by the Governor, under a penalty not exceeding £50 and not less than £10. Tidal waters within any areas so proclaimed are termed "Close Fisheries."

---

\* The restrictive and penal provisions in the Act did not come into force until the sixtieth day after its passing, *i.e.*, the 5th June, 1881.

† The mesh of a net is to be measured *diagonally*, and not from knot to knot along each side of the square, as is the rule in England and other countries.

# FISH AND FISHERIES.

### CLOSED FISHERIES.

Net-fishing of any kind is illegal in any tidal waters proclaimed by the Governor to be closed against such net-fishing, for the period specified by the Act.

### LICENSES.

A license for every fishing boat must be taken out annually—the license fee being £1; but, after the *thirtieth of June* the fee is reduced to 10s. for the remainder of the year.

Every fisherman must take out an annual license, subject to the like reduction for the broken period of the year; the fee is 10s., but subject to the same reduction.

### PRAWN-FISHING.

The riddling of prawns is only allowed while the prawns are alive.

In the river Hunter, and in any other river proclaimed by the Governor, an annual close season is or will be established. It extends from the *first of June* to the *thirtieth of September.*

### TORPEDOES, DYNAMITE, &c.

The use of *torpedoes, dynamite,* or *poisonous matter* in any waters is prohibited under a penalty not exceeding £40 nor less than £10. The prohibition, however, does not extend to persons lawfully authorized to explode torpedoes or dynamite.

### UNMARKETABLE FISH.

It is unlawful to sell, or expose for sale, or be in possession of any fish under the respective weights specified in the Second Schedule to the Act or prescribed by the Regulations. The prohibition does not, however, extend to collectors, owners of private fisheries, persons in possession of fish not intended for sale, or aboriginals, under certain conditions.

### OYSTER FISHERIES.

After validating the holdings, or premises of leases, acquired by certain persons under the "Oyster Beds Act of 1868," subject to their compliance with regulations for the management of the natural oyster-beds comprised in such holdings or promises, and forfeiture on default, the Act proceeds to the important subject of

### OYSTER LEASES.

Of these there are two descriptions:—

(1.) The first, *long leases* of shore and beds of tidal waters, may include an area not exceeding *25 acres.* The term must not exceed *thirty years,* or be less than *fourteen years.*

The *rent is 5s. per acre* for the *first four years,* and *£1 per acre* for the remainder of the term. No natural oyster-bed can be included in a lease, and no right is acquired by the lessee to the occupation of the shore (except so much as may be included in the area leased) unless for certain specified purposes.

No portion of the shore fronting private land can be included in a lease.

The further conditions and provisions, subject to which a *long term lease* for oyster culture is granted, are stated in section 28, and the penalties for injuring or interfering with oyster layings on such leases, in section 30. It should be observed that, before a lease can be granted, the notice of application for it must be published in the Gazette and some newspaper of local circulation. Objections on the ground of the area applied for containing a natural oyster-bed or part of such a bed may be lodged with the Commissioners within thirty days after the last published notice of application ; and an inspector may be sent to the locality to report as to the site of the lease applied for, and the validity of such objections. The lease may be revoked on default for two years by the lessee in taking proper measures to form oyster-beds or layings in terms of the lease.

(2.) *Annual Shore Lease for Oyster Culture.*—Under this form of lease any portion of the shore of a tidal water (not exempted or created a public oyster reserve under the Act) may be leased by the owner, lessee, or occupant of land abutting on the shore, at the rental of £1 for every 100 *yards* or portion thereof. The lease ranges from the margin of *mean high-water-mark* down to an imaginary line defining an average depth of *3 feet at low-water of spring tides*. The lease must be renewed annually, and the rent is payable in advance. It is, moreover, subject to such conditions and provisions as the regulations may prescribe. The lessee has the exclusive right during the currency of the lease (subject to the conditions and provisions just mentioned) of cultivating and protecting oysters within the limits of his lease, and of taking them for all purposes except burning for lime. He will, however, be required to mark out the position of his leased area by piles, stakes, or buoys, as directed by the regulations.

### Public Oyster Reserves.

The Governor in Council is empowered to declare any part of the shore abutting on any tidal waters, and any portion of the bed of any estuary, bay, lake, inlet, river, or creek influenced by the tides to be *exempt* from the leasing provisions of the Act, or to be a *Public Oyster Reserve*, subject to such regulations as may be framed under section 9.

### Royalty on Dredged Oysters.

Holders of dredging licenses dredging oysters on natural oyster-beds and layings belonging to the Crown, are required to pay for every reputed three-bushel bag of oysters a *royalty* not less than 1s. 6d. and not more than 4s. *per bag*. The royalty will vary for different rivers, and must be paid to the persons and in the manner directed by the Regulations.

### Oyster-dredging Licences.

Dredging for oysters on natural oyster-beds and layings abandoned or withdrawn from lease is prohibited, except to holders of oyster-dredging licenses. These licenses must be taken out either annually at a fee of £10, or quarterly at a fee of £3. The annual license bears date from the 1st January, and the quarterly license from the date of issue. In all cases these licenses are to be in accordance with the Regulations, and they authorize the *holder and his servants being the crew of any one " oyster-dredger"* to work on the appointed places and at the appointed times.

The penalty for unlawfully dredging for oysters is a sum not exceeding £20 and not less than £5, and the forfeiture of all oysters found in the offender's possession. Every dredger is required to produce his license on demand of an inspector or officer of police.

The Act also contains provisions for the registration of holders of dredging licenses, and for the marking of dredging boats.

## OYSTER-DEALERS' LICENSES.

A license is required to be taken out, at an annual fee of £5, for every person dealing in or selling oysters, whether wholesale or retail.

## MISCELLANEOUS PROVISIONS CONCERNING OYSTERS.

Further provisions are made by the Act, prohibiting the burning of oysters for lime—the dredging for oysters at any time between sunset and sunrise—providing for the carriage of oysters by sea in branded bags, and for their entry in the vessel's manifest—defining the property in oysters according to their position on leased land, private fisheries, or natural beds—and declaring the powers of the Governor in Council to absolutely close any oyster-beds against dredging for a term not exceeding three years.

## PRIVATE FISHERIES.

The private (or several) fisheries authorized to be established by the Act are limited to tidal-water fisheries, and have no applicability to inland waters uninfluenced by the tides, such as fresh-water lakes, and lagoons and rivers beyond the point where the effect of the tide is perceptible in affecting either the surface by way of rise and fall or the character of the water itself by imparting saline qualities. The admission of tidal water to private property, through or over Crown Lands (*i.e.*, foreshores) is the characteristic feature of the private fisheries contemplated by the Act; and this appropriation of what may be considered as one of the Crown's regalia, and the necessary interference with the shore and the *jus publicum* thereover, were considered as the circumstances which justified the legislation proposed.

The private fisheries authorized by the Act are established on application to the Minister* either by the person entitled to an estate of freehold in the land on which the fishing is to be exercised, or by the lessee with the consent in writing of his landlord or reversioner. The particulars required to be stated in the application, and the mode of dealing with it, are described in the 49th and 50th sections of the Act. If the application is granted, a license will be issued to the applicant on the payment by him of the sum of £10. The legal effect of such a license is to vest in the grantee, and all persons claiming under him, an exclusive right of stocking the fishery with fish of every description

---

\* The Colonial Secretary is the Minister charged with the administration of the Fisheries Act.

(including crustaceans and molluscs), and of taking all such fish so long as the license is in operation, *i.e.* so long as the terms and conditions subject to which it was granted shall be faithfully complied with. Moreover, the licensee will now be in a position to prosecute any persons who steal any such fish from his fishery, or who trespass within its limits. The grant empowers him to cut a trench or passage through the shore and to construct a sluice to admit the tidal water to his fishery, and any public rights-of-way or of navigation over his fishery, and of taking fish therein, are wholly abrogated by the grant. On the other hand, certain obligations are cast on the grantee or proprietor of the fishery. He will be required to construct and maintain substantial bridges of prescribed dimensions across any trenches or cuttings of the shore; also to mark the boundaries of the fishery by stakes or as prescribed by the regulations. A private fishery will pass as an incorporeal hereditament appurtenant to and with the ownership of the soil. Although not in terms granted as a franchise, it will more resemble a franchise than any other incorporeal hereditament. The licensee or grantee of the fishery will of course always remain liable to the Crown for the fulfilment of the conditions of the grant. The property in fish in a private fishery will vest in the owner of the fishery prior to, and not merely after, capture, as in the case of common of piscary.

### Limits of Jurisdiction.

The fourth section of the Fisheries Act, after declaring that the duty of protecting, developing, and regulating the public fisheries of the Colony shall be vested in the Commissioners appointed under the Act, enacts that the duties, powers, and authority of the Commissioners shall extend to the territorial limits of the Colony. What those limits are on the northern, western, and southern boundaries has been settled by statute, and there is therefore no doubt or difficulty in respect of the inland boundary-lines. By virtue of the Imperial Acts 14 Vic. c. 59, 18 and 19 Vic. c. 54, and the Order in Council thereunder of 6 June, 1859, validated by the next cited Act, 24 and 25 Vic. c. 44, and an Order in Council of 9 August, 1872,* the territory of New South Wales is defined on the northern boundary by a line extending from Point Danger, a short distance to the northward of the Tweed River, to the head waters of the Macintyre River, where the 29th parallel of south latitude impinges on that river; thence westward by that parallel to its intersection by the 141st meridian of east longitude; southward, along that meridian to the waters of the river Murray; and eastward by the southern bank of that river to a point where a straight line from Cape Howe touches the nearest source of that river. By the 5th section of 18 and 19 Vic., c. 54, it is declared "That the whole watercourse of the said river Murray, from its source therein described to the eastern boundary of the Colony of South Australia, is and shall be within the territory of New South Wales : Provided nevertheless, that it shall be lawful for the Legislatures, and for the proper officers of Customs of both of the said Colonies of New South Wales and Victoria, to make regulations for the levying of Customs Duties on articles imported into the said two Colonies respectively by way of the river Murray, and for the

* See Despatch relating to Pental Island. *Votes and Proceedings*, 1872-3, p. 519.

punishment of offences against the Customs Laws of the said two Colonies respectively committed on the said river, and for the regulation of the navigation of the said river by vessels belonging to the said two Colonies respectively : Provided also that it shall be competent for the Legislatures of the said two Colonies, by laws passed in concurrence with each other, to define in any different manner the boundary-line of the said two Colonies along the course of the river Murray, and to alter the other provisions of this section." By virture of the Order in Council already referred to, Pental Island, on the river Murray, was declared to belong to Victoria, and not to New South Wales. Until, therefore, by the concurrent legislation of the Colonies concerned, the river boundary-line of this Colony is altered, the channel of the river Murray from the intersection of that river by the 141st meridian belongs to New South Wales, and, of course, with the channel or watercourse of the river, all incident powers of legislation, and all territorial jurisdiction in and over the water and soil of the river must be considered to have passed to and become vested in New South Wales, subject to the express provisoes declared by the Act.

It only remains to consider what is the seaward limit of the jurisdiction on our eastern seaboard and on the coasts of our dependencies. It may be considered as a rule of law firmly established on the authority both of publicists and decided cases, that the portion of the sea washing the coast of a maritime State which lies within the range of cannon-shot from land, is the territorial property of that State, and subject to its municipal jurisdiction, but that beyond that limit and out of the reach of cannon-shot, " common or universal use" (to use the words of Lord Stowell), is presumed. The sea within this limit of cannon-shot, or as the limit has in modern times been expressed—one marine league or 3 miles—is considered to belong to every independent maritime State.

### FISHES PROTECTED BY THE ACT.

Schnapper, black bream, silver bream, black fish, black rock cod, red rock cod, Gurnets, flatheads, mullets (including Flat-tail and sand mullets), whiting, flounders, soles, pike, trevally, garfish, crayfish, Murray cod, and Murray perch.

### CORMORANTS OR SHAGS.

As these birds are found to abound on the coast, and to be most destructive to fish, regulations are to be framed by the Commissioners for the payment of rewards to those who kill them, but the sum must not exceed 1s. per bird. The following short description and nomenclature of those species which are ordinarily found on our seaboard and rivers has been supplied by the Curator of the Australian Museum, Mr. E. P. Ramsay, F.L.S. :—

1. Great Black Shag (*Graculus novæ hollandiæ*). Black all over ; bill and skin round face, black ; sometimes when breeding a white spot on the thighs and a few white striæ on the neck.
2. Small Black Shag (*Graculus stictocephalus*). Glossy black all over ; bill and skin round face, black ; eye, green.
3. Small Black and White Shag (*Graculus melanoleucus*). Sides of the head and all under side, white ; top of the head and all upper parts, black ; skin round face, and the bill, yellow.

4. Large Black and White Shag (*Graculus leucogaster*). Throat and all under surface, white; all the upper parts, black; bill and skin round face, black; eye, green.

5. Blue-eyed Shag, or Southern White-bellied Shag (*Graculus varius*). Sides of face and all under surface, white; head and all upper surface, black; skin round the eye, bright blue, with a spot of bright yellow in front of the eye; bill, brown.

LEASES UNDER THE OYSTER-BEDS ACT OF 1868.

| Locality. | Date of Expiry of Lease. |
|---|---|
| 121 acres Oyster Creek, Clarence River | 31 Dec., 1881 |
| 98 acres ,, ,, ,, ,, | 31 Dec., 1881 |
| 106 acres ,, ,, ,, ,, | 31 Dec., 1881 |
| Clarence River | 1 April, 1883 |
| Camden Haven | 1 May, 1883 |
| Wallis Lake (Cape Hawke) | 1 Sept., 1883 |
| Port Stephens | 1 May, 1883 |
| Broken Bay | 1 Oct., 1883 |
| Kissing Point, Parramatta River (1 acre) | 1 April, 1884 |
| From the Embouchure of George's River to Botany Heads | 1 Jan., 1884 |
| George's River | 1 April, 1883 |
| Shoalhaven and Crookhaven River | 1 Sept., 1883 |
| Clyde River and Tributaries | 1 Sept., 1883 |
| Tuross River | 1 March, 1884 |

NOTE.—It is believed that some few of these leases were always not valid as comprising natural oyster beds. The majority, however, did comprise natural oyster beds.

According to the Second Schedule of the Act, the following table for the lawful weight or minimum at which the undermentioned fishes may be sold:—

*Lawful Weights for Fish.*

| | Description of Fish. | Weight in ounces avoirdupois. |
|---|---|---|
| Marine | Schnapper or Red Bream | 16 ounces. |
| | Bream (Black) | 8 ,, |
| | ,, (Silver) | 4 ,, |
| | Blackfish | 8 ,, |
| | Rock-cod (Black or Red) | 8 ,, |
| | Gurnet | 4 ,, |
| | Flathead | 8 ,, |
| | Mullet— | |
| | Sea [including the variety commonly known as hard-gut mullet] | 12 ,, |
| | Flat-tail | 4 ,, |
| | Sand | 4 ,, |
| | Whiting | 4 ,, |
| | Flounder | 4 ,, |
| | Sole | 4 ,, |
| | Pike | 8 ,, |
| | Trevally | 8 ,, |
| | Garfish | 3 ,, |
| | Lobster (or Crayfish) | 16 ,, |
| Fresh-water | Cod (or Murray Cod) | 16 ,, |
| | Perch | 4 ,, |

## FISH AND FISHERIES.

## THIRD SCHEDULE.

| Name of River or Place. | Date of expiry of Proclamation closing Oyster-beds. |
|---|---|
| Bellinger River | 29 July, 1881. |
| Bermagui River | 26 August, 1881. |
| Brow Lake | 16 August, 1881. |
| *Candlegut Creek | 7 April, 1881. |
| *Congo Creek | 7 April, 1881. |
| Cutagee Creek | 31 May, 1881. |
| *Dunn's Creek (Hunter River) | 28 April, 1881. |
| Hunter River | 25 November, 1881. |
| Jervis Bay | 26 August, 1881. |
| Lake Macquarie | 31 May, 1881. |
| Manning River | 19 January, 1882. |
| Nambucera River | 26 August, 1881. |
| Narrawillie Creek | 12 January, 1882. |
| Narrabeen Lagoon | 20 May, 1881. |
| Panbula River | 26 August, 1881. |
| Port Macquarie | 4 February, 1882. |
| Richmond River | 26 August, 1881. |
| Salt Water Creek (near Manning River) | 28 September, 1881. |
| *Smith's Creek (Hunter River) | 28 April, 1881. |
| Tilba Tilba Lake | 26 August, 1881. |
| Tomago River | 26 August, 1881. |
| Towamba Lake | 26 August, 1881. |
| Tweed River | 26 August, 1881. |
| Twofold Bay | 26 August, 1881. |
| Wogongo River | 26 August, 1881. |
| Port Jackson, including Middle Harbour and Lane Cove and Parramatta Rivers and their inlets | 29 July, 1881. |

* The date on which the Proclamations closing these waters respectively expired were extended by notification in the Gazette for the respective periods following, viz. :—

    Candlegut Creek—For a period of twelve months from 8th April, 1881.

    Congo Creek—For a like period of twelve months.

    Smith's Creek ⎱ Each for a further period of six months from 28th April,
    Dunn's Creek ⎰ 1881.

*Note.*—In addition to the above proclaimed closures, *Port Hacking* was closed for twelve months, from the 8th April, 1881.

On the 6th June, 1881, the Governor, by Proclamation, closed the tidal waters of Tuggerah Beach Lakes, including Manmurra and Budgewi, and Narrabeen and Deewhy Lagoons against the use of fishing-nets for the period of two years from that date.

## SPECIAL REGULATIONS

*Relating to the Oyster-bed Leases validated by sec. 27 of the Fisheries Act,* 1881.

In these regulations the term "lease" means a lease or promise of a lease made under the Oyster-beds Act of 1868 as validated by sec. 27 of the Fisheries Act 1881. "Lessee" means the holder of or person entitled for the time being to any such lease. "Inspector" means any Inspector or Assistant Inspector. "Bed" means any natural oyster-bed as defined by the said Oyster-beds Act of 1868, or any other laying included within any such lease.

2. The yearly rent payable in respect of all oyster-beds under lease or promise thereof shall continue to be paid in advance as heretofore, to the Colonial Treasurer or officer authorized by him to receive the same. And on default of payment of any such rent for one calendar month after the same shall have become due the interest of the lessee so in default shall be declared forfeited to Her Majesty, and upon notification of such forfeiture in the Gazette such interest shall be forfeited and the lease or promise thereof shall thereupon terminate accordingly.

3. Every bed shall be open to inspection by an Inspector, at any time between sunrise and sunset, and he may test, by dredging or in any other way, the condition of such bed and of the oysters thereon.

4. No oysters capable of being passed through a metal ring having a diameter of one inch and three-quarter (inside measurement) shall be taken from any bed by the lessee or any other person, except for the purpose of being laid down for culture in some other portion of the area under lease.

5. On a report in writing of the Inspector, furnished to the Commissioners, that the oysters in any specified bed or portion of a lease are out of condition, whether by reason of spatting or for any other cause, the Commissioners may recommend the Governor to prohibit the removal of any oysters from such bed. Such prohibition shall be notified in the Gazette, and a notice thereof served on the lessee, or posted to or delivered at his usual place of business shall be sufficient notice to the lessee that the taking of oysters from such bed is prohibited during the period named therein.

6. The lessee and every person employed by him for the purpose of dredging or taking oysters from a bed under lease shall each be severally liable, for any breach of these regulations or non-compliance therewith, to a penalty not exceeding £20 for every such offence or default.

7. Upon a second or subsequent conviction of the lessee for any offence under these regulations, his lease, as validated by section 27 of the said Act, may on the recommendation of the Commissioners, and in addition to any penalty prescribed by these regulations, be declared by the Governor, with the advice of the Executive Council, to be forfeited to Her Majesty.

8. Upon receipt of a report by an Inspector that any bed has been so stripped of oysters or otherwise so mismanaged by the lessee during his tenancy that the production of oysters on such bed has, in such

# FISH AND FISHERIES. 171

Inspector's opinion, been so reduced as to threaten the destruction of the bed, or to render it unfit to be dredged under Royalty, or to be worked, upon the determination of the lease, the Commissioners may call upon the lessee to show cause why they should not recommend the Governor, with the advice aforesaid, to declare the lease of such lessee to be forfeited. And upon the receipt of a recommendation by the Commissioners recommending the forfeiture of any such lease, the Governor may, by notification in the Gazette, declare the same to be, and the same shall thereupon be, forfeited accordingly.

## GENERAL REGULATIONS.*

Colonial Secretary's Office, Sydney, 30th June, 1881.

HIS Excellency the Governor, with the advice of the Executive Council, has been pleased, in accordance with section 9 of the Fisheries Act, 1881, to make the following Regulations for giving effect to the provisions of that Act.

HENRY PARKES.

### Conduct of Business—Duties of Officers.

1. The Commissioners will meet for the transaction of business at their office on every Monday and Thursday at 2 o'clock. If either of these days be a public holiday, the meeting will be held on the day following.

2. The President may, by circular to be addressed by the Secretary to the residence or office of each Commissioner, convene a special meeting of the Commission on any day and hour mentioned in such circular.

3. The common seal of the Commissioners shall not be affixed to any document, paper, or writing, except by direction of the President, or, in his absence, of the Chairman, and in the presence of some Commissioner.

4. It shall be the duty of the Secretary to prepare the business paper for each meeting of the Commission, to take and record the minutes of proceedings at the same ; to conduct all correspondence, and keep all such books of account, vouchers, reports, documents, plans, and charts as the Commissioners may direct or require ; to keep the common seal of the Commissioners, and to affix the same to any document or paper if so directed by the President or Chairman ; to give such instructions to Inspectors, Assistant Inspectors, and other officers and persons appointed under the Fisheries Act as the Commissioners shall direct, or as the regulations may prescribe ; and generally to fulfil zealously and to the best of his ability all duties and obey all directions imposed on or given to him by the Commissioners.

5. Every Inspector of a Division must report in detail to the Commissioners once in every month, or oftener if practicable, as to the state of the Fisheries included within his Division so far as he has been able by personal inspection or from trustworthy information to ascertain the

---

\* See Government Gazette of June 30, 1881.

same. He will be required to report especially as to the condition of the natural oyster-beds and deposits of oysters in all waters within his Division; whether the oysters are in marketable condition or otherwise, and what number of bags of oysters have been dredged from such beds during the preceding month, together with the amount of the royalty paid or required to be paid thereon; to receive and transmit to the Commissioners all applications for leases which may be sent to him; to fill up, to the best of his knowledge and judgment, and to transmit without delay to the Secretary, all forms of returns issued to him; to inform the Secretary of all breaches of the Act or the Regulations, whether by acts of commission or default, and to await the instructions of the Commissioners before laying an information against offenders or defaulters; to take particular notice of the movements and habits of the various kinds of useful fish, whether included in the Schedules of the Act or not; to report to the Secretary the existence of any source of pollution to the waters under his inspection, or of any mortality, disease, or ill-condition of fish, oysters, lobsters, crabs, or prawns; to ensure cleanliness and good order in his boat and crew (if any); to keep true and particular accounts of all expenses incurred by him when absent from his post on inspection or other service, and to enter the same in his journal; to keep a diary or journal in which he shall enter every day's work, and any other matter affecting the state of the fisheries in his Division; to visit and report upon all stations within his Division at least once in each quarter, unless otherwise directed by the Commissioners; to report himself at the office of the Commissioners whenever in Sydney on duty; to promote, by every means in his power, good feeling and concert of action in his relations with all Inspectors, Assistant Inspectors, and other persons concerned with the the administration of the Act within his Division, and generally to devote his best energies to the administration of the Act and Regulations, and to the performance of all duties delegated to him by the Commissioners.

6. The duties of the Assistant Inspector shall be the same *(mutatis mutandis)* as those of an Inspector of a Division, except that his reports shall be in duplicate, one to be transmitted to the Inspector of the Division, the other to the Secretary of the Commissioners, and that he will not, unless required by the Commissioners, visit or inspect any fisheries except those included within his district.

7. The duties of an Acting Assistant Inspector shall be to assist the Inspectors so far as may be in his power; to inform them of any breach of, or non-compliance with, the Fisheries Act or the Regulations, and, in the absence of an Inspector, to report any such breach or non-compliance to the Secretary, and receive the instructions of the Commissioners thereupon; to take all legal and other proceedings which he may be instructed to take; to protect the Revenue by all means in his power from being defrauded by non-payment of royalties or otherwise; to report all cases of oysters being found in a vessel in unbranded bags or otherwise in contravention of the Act or Regulations; to inform Inspectors of all matters relating to the fisheries at or near his station which comes to his knowledge; and to carry out all instructions received from the Commissioners to the best of his ability.

FISH AND FISHERIES. 173

*Marking of Licensed Fishing Boats.*

8. The owner or person in charge of a Fishing Boat licensed under section 12 shall paint and keep painted on the inside of the truck or transom of such boat the Christian and surname of such owner, and the words "Licensed Fishing Boat" in legible Roman letters, not less than 3 inches in length.

*Fishing Boat Licenses.*

9. Fishing Boat Licenses under section 19 shall be in form A hereto. The fee of £1 for such license shall be paid either to the proper officer at the Treasury, Sydney, or to the Police Magistrate, or Clerk of Petty Sessions of the Bench nearest to the fishery wherein such license is intended to be exercised. Owners of boats fishing within the limits of the Home Fisheries (*i.e.* between Port Stephens and St. George's Basin, south of Jervis Bay), should, wherever practicable, take out licenses and pay license fees at the Treasury, at Sydney. In the other Divisions the licenses must be taken out and the fees paid at the nearest Court of Petty Sessions. The license must be renewed in like manner annually at the like fee of £1. After the 30th June in any year the fee for the broken portion of the year will be 10s.

*Fishermen's Licenses.*

10. Fishermen's Licenses under section 20 shall be in form B hereto. The fee of 10s. for such license shall be paid at the like places, and to the like persons, as Fishing Boat License fees. The license must be renewed annually in like manner. After the 30th June in any year the fee for the broken portion of the year will be 5s.

*Oyster-dredging Licenses.*

11. Oyster-dredging Licenses under section 36 shall be in the form C hereto. They may be taken out either for the whole year or for any quarter of a year. The fee for the yearly license is £10, for the quarterly is £3, to be taken out and paid in each case at the Court of Petty Sessions nearest to the place where the license is to be exercised or (at the applicant's choice) at the Treasury in Sydney. The Police Magistrate or Clerk of Petty Sessions is the proper officer (out of Sydney) to whom application for licenses should be made.

*Oyster Dealers' Licenses.*

12. Oyster Dealers' Licenses under section 42 shall be in the form D hereto, and applications for such licenses must be made either to the Secretary of the Commissioners, or to the Police Magistrates of the nearest Bench. The license fee of £5, or £2 10s. for the broken portion of the year, if the application be made after the 30th June in any year, must be paid to the proper officer at the Treasury in Sydney. The icense must be renewed annually by application made in like manner.

*As to Transmitting and Accounting for Moneys.*

13. The General instructions to Public Officers issued by the Treasury shall govern all Police Magistrates, Clerks of Petty Sessions, and all other officers collecting or receiving License Fees or other moneys under the Act or the Regulations.

### Testing Length and Mesh of Nets.

14. Every net for the purpose of testing the length thereof shall be measured along the cork line or line on which such net is hung. The size of mesh in every case shall be ascertained by measuring the length on the diagonal, or between knot and knot of opposite corners, the net being first wetted and being tanned, barked, or otherwise prepared for use. In case of dispute or doubt, a ½ pound weight shall be slung or attached to one knot of a mesh, in order to produce a fair strain or extension, and the space between the knots shall be measured forthwith while the mesh remains extended If the net to be measured is dry, the part to be measured shall be soaked either in fresh or salt water for not less than ten minutes, and the mesh so soaked shall then be measured.

### Marketable Prawns.

15. Marketable prawns are prawns not caught during the close season prescribed by section 16 for certain fisheries, and being not less than one inch and a half in length, measured from a point between the eyes to the end of the tail ; and all smaller prawns shall be deemed to be unmarketable, as well as all prawns (whatever the size) caught during such close season as aforesaid : Provided that if in any basket of prawns, or other vessel or receptacle in or upon which prawns shall be exhibited for sale, a number of unmarketable prawns not being on the whole more than one-tenth part of the contents of such basket, vessel, or receptacle, shall be found so exhibited, no liability under the Act or Regulations shall be incurred by any person in respect of the exhibition of such prawns.

### Priority among Netters.

16. The right of first shooting and hauling a net on any fishing-ground shall belong to the licensed fisherman who first arrived on the ground with his boat and net ready for shooting ; and the next turn shall belong to the licensed fisherman who arrived next after such first-mentioned person, and so on in order of arrival. An unlicensed fisherman shall not be entitled to shoot a net on any fishing-ground until every licensed fisherman then being on the ground with boat and gear ready for shooting shall have had his turn. This regulation shall not apply to prawn fishermen. Two or more turns may be taken at the same time if the water to be fished permits of double-banking, but no net shall be shot round an inner net within a boat's length of the cork line of such net.

### Turns in Oyster-dredging.

17. Every licensed oyster-dredger shall take his turn and place on the bed according to the time of bespeaking such turn, unless the Inspector chooses to permit several boats to work at the same time on the same bed, in which case the dredgers shall work in obedience to the Inspector's directions. All turns must be bespoken from the Inspector at least twenty-four hours in advance. Every dredger bespeaking a turn shall lose it if he is not on the ground with boat and gear ready punctually at the time appointed by the Inspector.

# FISH AND FISHERIES. 175

### Inspection of Artificial Oyster-beds and Layings.

18. Artificial oyster-beds and layings of any lessee (not being private fisheries under Part III) may be inspected by any Inspector once in every month, or oftener if directed by the Commissioners, upon such Inspector giving twenty-four hours' notice either to the lessee or his manager or his agent in charge of the beds of the intention to inspect the same. In the course and for the purposes of such inspection the lessee, his manager, agent and servants shall, at the request of the Inspector, test the state of any bed or laying required by the Inspector to be tested, by dredging or gathering by tongs, hand or otherwise (according to the nature of the bed) oysters from such bed, under a penalty upon every such person refusing or neglecting to do so of £5.

### Dredging on Natural Oyster-beds.

19. Natural oyster-beds shall not be dredged, nor shall any oysters be taken from such beds (except by an Inspector in the performance of his duties) unless the same shall have been notified in the Gazette to be open for dredging or otherwise than in accordance with the directions of the Inspector of the fishery within which such oyster-beds are situated.

### Cleaning, &c., of Natural Beds.

20. The Inspector may, notwithstanding any such notification as aforesaid, close any natural oyster-bed, or portion thereof, for the purpose of cleaning, levelling, or otherwise improving the same, in any case where and for such time as in his opinion such closing shall be necessary; but such closing shall be subject to revocation and modification by the Commissioners. No dredging shall take place on a bed or portion so closed until the same shall be declared by the Inspector or the Commissioners to be again open for dredging.

### Inspector may in certain cases stop Dredging.

21. Whenever the Inspector shall be of opinion that the further dredging of any bed would be injurious, or that the oysters therein are out of season, and not fit for food, whether by reason of spatting, freshets, or any other cause, he may prohibit the dredging or taking of oysters from such bed for such period, subject to the directions of the Commissioners, as he shall think necessary, and during such period no person shall dredge for or take oysters from any such bed under a penalty not exceeding £20.

### None but Licensed Dredgers to Dredge.

22. No persons shall be allowed to dredge for or take oysters from any natural oyster-bed who shall not, on demand of the Inspector, produce the oyster-dredging license for inspection.

### No turn allowed unless to Licensed Oyster-dredgers.

23. Every person bespeaking a turn for dredging on any such bed must produce his license (if required by the Inspector) before such turn shall be allowed.

*Scale of Royalties on Oysters Dredged from Natural Oyster-beds.*

24. The sums payable by way of royalty on oysters dredged from natural oyster-beds shall for every bag containing or reputed to contain three bushels be payable according to the following scale :—

| River, bay, inlet, or other locality where or near to which the natural beds are situated. | Amount of royalty per bag. |
|---|---|
| | s. d. |
| The Tweed | 2 0 |
| The Richmond | 3 0 |
| The Bellinger | 3 0 |
| The Nambuccra | 2 0 |
| The Brunswick | 2 0 |
| Port Macquarie | 2 6 |
| The Manning | 2 6 |
| The Hunter | 4 0 |
| Port Hacking | 3 0 |
| Jervis Bay | 2 6 |
| Twofold Bay | 3 0 |
| Any place other than those above specified | 2 6 |

The above scale of royalties shall be in force until the Commissioners otherwise appoint, but shall not apply to any rivers or other oyster-bearing waters mentioned in the Third Schedule to the Act until the proclamation closing the same shall be rescinded by a proclamation published in the Gazette, or the term of closure limited by such Proclamation shall have expired.

*Royalty, how payable.*

25. Royalties according to the prescribed scale for the time being shall be paid to the Inspector, or in his absence, to the Assistant Inspector, on the dredging or taking of oysters from the bed, and before the same are removed or shipped on board any vessel; but in all cases where such oysters are consigned to dealers or persons resident at any port or place within the Colony, such Inspector may give the consignor a shipping permit in the prescribed form, and issue a duplicate thereof for the master of the vessel in which the oysters are intended to be shipped. Such permit shall state the number of bags shipped, the royalty payable thereon, the brands and marks of the bags containing the oysters shipped, and the names of the consignor and consignee. On the vessel's arrival at the port of destination, the master shall deliver the said permit to the proper officer of Customs, and the consignee shall on payment to such officer at the Custom-house, or to the proper officer of Customs, of the amount stated in the permit to be due as royalty on such oysters, be entitled to take delivery thereof. If the royalty be not paid within twenty-four hours after the vessel is reported to the Customs, the proper officer of Custom shall take possession of the oysters, and either warehouse the same or cause them to be sold as the Commissioners shall direct. In case of sale the whole proceeds of sale shall be paid into the Treasury, and after the amount due as royalty shall have been deducted therefrom, the balance shall be paid to the consignee or his authorized agent.

## FISH AND FISHERIES.

### Disposal of Cultch.

26. All dead shells and cultch, whether with or without young oysters attached thereto, dredged up by any person dredging for oysters or natural oyster-beds, shall be thrown back by such person in such place as the Inspector shall appoint; and all oysters below marketable size so dredged up shall be returned to the beds by the person who dredged them up, at such places and at such times as the Inspector shall direct.

### Marketable oysters.

27. No oysters which can be passed through a metal ring having a clear inside diameter of one inch and three-quarters shall be deemed to be marketable oysters; and it shall be unlawful to dredge for, take, consign, or expose for sale any such under-sized oysters, but they may be taken for the purpose of being laid down on another bed or in different water. This regulation shall apply to all lessees of Crown Lands for oyster culture, and to all grantees of private fisheries.

If any person exposes for sale in any shop, boat, vehicle, stand, or place, any oysters not marketable withing the meaning of this regulation, he shall be liable to a penalty not exceeding £10; but this regulation shall not apply to persons selling oyster-spat or brood to lessees and owners of private fisheries, otherwise than for consumption as food, or to oysters imported into this Colony from another Colony.

### Permit to Lessees and others to procure Spat from Crown Lands.

28. A lessee or an owner of a private fishery desirous of obtaining a supply of spat or young oysters for the purpose of stocking his leased land or private fisheries, must apply to the Commissioners for a permit, which will be granted on the terms and conditions therein specified, if the Commissioners are of opinion, after receiving the report of the Inspector, that there is sufficient young oyster stock on the foreshores of the tidal waters, on or near to which is situated the lease or private fishery for which the supply is required, to allow of its distribution among lessees and owners of private fisheries. Such permit shall describe the area within which, and the time during which the permission may be exercised. The fee payable for such permit shall be £1, and shall be paid to the Colonial Treasurer or officer authorized by him to receive fees under the Act.

### Disposal of seized Oysters.

29. Oysters seized under the authority of the Act shall be taken charge of by the seizing officer and delivered to the Inspector or person authorized by the Inspector to take delivery of such oysters. If the Inspector shall be of opinion that the oysters so seized will die or become unmarketable before adjudication, he shall cause them to be sold to the best advantage, either by private contract or by public auction at his discretion; and the proceeds of sale shall be paid into the Treasury to the credit of a special Suspense Account. If the oysters seized shall afterwards be adjudged to be forfeited, such proceeds shall be paid into the General Revenue; if not, they shall be paid to the person entitled to the oysters. But if the person in possession of the oysters at the

time of seizure wishes them to be retained in the custody of the seizing officer or Inspector, such officer shall so retain them, but at the risk of such person as aforesaid, until the case has been disposed of, either by adjudication of the Justices, or by order of the Commissioners directing the Inspector to abstain from any or to stay further proceedings. Notwithstanding anything contained in this Regulation, the Commissioners may direct the seizing officer or Inspector to dispose of any oysters so seized as aforesaid in such way, and subject to such terms and conditions as they may by writing addressed to such officer appoint.

### *Branding of Oyster-bags, &c.*

30. Oysters shipped on board any vessel shall be placed in bags branded with the Christian and surname of the consignor and consignee and with the name of the river or place where such oysters have been dredged or taken. For example :—Oysters dredged by John Smith at the Clarence, and shipped there for Sydney, must be placed in bags branded JOHN SMITH, CLARENCE RIVER (the name and address of the consignor), and ROBERT BROWN, SYDNEY (the name and address of the consignee). All such brands shall be in Roman capital letters not less than 3 inches in length, and shall be placed on the outside of the bag, at or near the middle thereof.

31. No brand of a consignor's name and address shall be affixed to any bag containing oysters so shipped unless the same has been registered either by the Inspector at the place where such oysters were dredged or taken, or with the Inspector of the Division within which the same were dredged or taken, or at the Office of the Commissioners of Fisheries in Sydney, and unless a certificate of registration under the hand of such Inspector, or a like certificate under the hand of the Secretary to the said Commissioners, shall have been given to the person desiring to register such brand.

32. The fee on the registration of an oyster-bag brand shall be two shillings and sixpence, which must be paid before the issue of the certificate to the registering officer.

### *Rewards on Destruction of Cormorants or Shags.*

33. On the production to an Inspector of any number of heads of cormorants or shags, not being less than half-a-score, such Inspector shall give the person producing the same a receipt therefor, and on production of such receipt at the Treasury in Sydney, the person producing the same shall be entitled to receive a sum equal to sixpence for each head (not being in the whole less than ten) of the larger cormorant or black shag, and equal to fourpence for each head (not being less than aforesaid) of the smaller cormorant or white-bellied shag.

34. The Inspector receiving such heads shall forthwith after recording the same, destroy them by fire in the presence of some Justice of the Peace, Officer of Customs, or of the Department of the Marine Board, or Constable, who shall certify the fact of such destruction under his hand on the butt of the receipt book of the Inspector.

# FISH AND FISHERIES.

### As to Public Oyster Reserves.

35. Portions of Crown Lands declared exempt from lease under the 33rd section, and notified in the Gazette as Public Oyster Reserves, shall be divided into two classes :—(1.) Public Oyster Reserves for recreation. (2.) Public Oyster Reserves for oyster breeding purposes.

### Reserves to be under Control of Inspectors.

36. All public oyster reserves shall, subject to the directions of the Commissioners, be under the control and supervision of the Inspector of the tidal waters wherein such reserves are situated. And such Inspector, or any officer of police, or Constable, may apprehend and lodge in custody any person found removing oysters from any Recreation Reserve in a bag or other vessel or receptacle other than a bottle, and for his own consumption, or found wantonly destroying any oysters on such reserve, or conducting himself thereon in a disorderly manner, using profane, obscene, or disgusting language, drowning or destroying dogs, goats, cats, or any animal whatsoever, or depositing any dead carcase on, or within one hundred yards of, such reserve; or exposing his person, or annoying the residents or passers by. And any person charged with the commission of any such act as aforesaid shall, on conviction thereof, forfeit and pay a sum not exceeding £10.

### Licenses to gather Oysters and Spat for sale or culture from Breeding Reserves.

37. Annual licenses may be granted on application to the Commissioners or to the Police Magistrate or Clerk of Petty Sessions of the Bench nearest to a public oyster reserve for oyster-breeding purposes, to any persons to gather oysters and spat therefrom for the sale or culture, whether for laying down on the layings of the applicants or on those of other persons. The fee for every such license shall be five shillings; but each license shall authorize only the person named therein to gatser such oyster or spat. If no particular portion of the reserve is specified in the license, the licensee can gather throughout the entire reserve; but **if a particular portion be specified, the licensee will be restricted to that portion.** Notwithstanding the grant of licenses, a public oyster reserve for breeding purposes, or any portion thereof, may, if the Commissioners think fit, be withdrawn from use as a gathering ground for such time as they may appoint. And all licenses shall be deemed to be granted subject to the conditions af such withdrawal.

### Marking of Public Oyster Reserves.

38. All public oyster reserves, and all portions of Crown Lands exempted from lease under section 33 of the Act, will be marked or defined by stakes, buoys, or such other mode as the Commissioners shall direct; but such marking or definition shall not be deemed to be mandatory in any case where the boundaries of any such reserve have been defined in the Gazette notifying such reserve.

### As to Recognizances under Section 62.

39. The gaoler or person in charge of any gaol, lock-up, or police station, may liberate any person lodged in his custody under the 62nd section of the Act, on such person entering into recognizance, with or

without sureties, as such gaoler or person in charge as aforesaid shall think fit, conditioned that he will appear for examination before a Justice of the Peace at a place and time to be therein specified. And such recognizance shall be of equal obligation on the parties entering into the same, and be liable to the same proceedings for estreatment as if the same had been taken before a Justice of the Peace. And such gaoler or person in charge as aforesaid shall make the like entries and take the same proceedings (as shall also the Justice before whom the same are tried, and the Clerk of the Peace and all other persons) in respect thereto, as are required or directed by the Towns Police Act, 2 Vic. No. 2, in respect of persons charged with any petty misdemeanour thereunder.

### Regulations for the Hauling and Landing of Gar-fish and Prawn Nets.

40. Every *bonâ-fide* gar-fish net within the meaning of section 11 of the Act, and every net having a mesh in the centre less than 2 inches, and in the wings less than 3 inches, shall be emptied in the water, and shall not be hauled ashore to a beach or strand.

41. Whenever a net used for catching prawns shall be hauled ashore to a beach, one end of such net shall be opened so as to allow all undersized fish to escape, and only the flue or centre of the net shall be brought ashore.

### Marking Oyster-beds and Leased Areas.

42. The position of natural oyster-beds shall be marked by piles or stakes driven into the bottom at such places and in such manner as to define with reasonable accuracy the length, width, and shape of the beds. Such piles or stakes shall show above the line of high-water-mark at least 6 feet, and shall be painted red, with the words, "Oyster-bed, No. " [*stating the number of the bed on the particular river or tidal water*] painted in white Roman letters at least 3 inches in length, on a black cross bar or batten, near the head of the pile, or stake. Piles must be not less in average diameter than 6 inches. Buoys of iron or wood may be substituted for piles or stakes, if the water be too deep for piles or stakes, or in cases of necessity, or temporarily, at the discretion of the Inspector.

43. All areas leased under section 28 or 32 shall be marked at each corner by white piles of the same dimensions, and showing out of water to the same height as in the case of natural oyster-beds, or by iron or wooden buoys, in each case painted white, at the option of the lessee, if the bottom at any such corner shall be rock, or otherwise incapable of holding a pile, or if the water be deeper than 2 fathoms at low water. In all cases the words "Oyster-lease, No. " *(inserting the registered number)* shall be painted in black Roman letters at least 3 inches in length, on a white cross-bar or batten near the head of the pile or stake, or across the face of the buoy.

44. Wherever practicable the piles, stakes, or buoys shall be so placed, both on natural oyster-beds and on leased areas, that imaginary lines connecting the boundaries across the river, creek, or inlet shall be at right-angles to the direction of the tide.

45. If two or more contiguous portions leased under section 28 or 32, or both, be held by the same lessee, the outer limits only of the block need to be marked by piles, stakes, or buoys, as directed by these Regulations, unless the Inspector satisfies the Commissioners that all the corners or limits of each such portion ought to be marked.

46. Every lessee who shall neglect to mark and keep marked the limits of his leased area as required by these Regulations, shall be liable to a penalty not exceeding £10.

47. Any person who destroys, removes, or injures any pile, stake, or buoy, marking out the position of a natural oyster-bed, or of any leased area, or who defaces or obliterates any letters on any such pile, stake, or buoy, shall be liable to a penalty not exceeding £20 for every such offence.

48. No person engaged in dredging oysters on a natural oyster-bed shall make fast his warp to any pile, stake, or buoy defining such bed under a penalty for each offence not exceeding £2.

## Disposal of forfeited Fish.

49. Fish ordered to be forfeited to Her Majesty, pursuant to the provisions of section 14 or 18, may, at the discretion of the Justice or Justices as the case may be, be handed over to the authorities of the Benevolent Asylum, or of any hospital or charitable institution nearest to the place where such forfeited fish are; but, if such fish are unfit for food, such Justice or Justices shall order them to be destroyed.

## Disposal of forfeited Oysters.

50. Oysters forfeited to the use of Her Majesty, pursuant to the provisions of section 39 or 46, must be destroyed if unfit for food, but otherwise shall be sold by auction, unless the Commissioners direct that they shall be laid down on a natural oyster-bed or in some tidal water. If sold, the proceeds of sale shall be paid into the General Revenue.

## Disposal of forfeited Nets.

51. Fishing nets forfeited to the use of Her Majesty pursuant to the provisions of section 18 shall be sold by auction, and the proceeds of sale shall be paid into the General Revenue ; but such sale shall not take place until the expiration of sixty days from the date of the conviction under the said section. The Commissioners however may in any case, if they shall be of opinion that the offence or default leading to the forfeiture was committed or arose through inadvertence or mistake, recommend the Governor to restore any forfeited net to its owner.

## Penalties on Breach of Regulations.

52. Any person who shall commit any act in breach of, or be guilty of any default or non-compliance with the requirements or prohibitions of any of the foregoing regulations, shall in every case where no penalty or forfeiture has been in such case provided be liable to a penalty not exceeding £10.

## CHAPTER XV.
### Index of Local Names.

In the following list only those local names are given which are used in Australia, and such as from their importance or wide-spread use have a special interest. The scientific name is always that of Günther, in the "Study of Fishes," and Macleay's list. The letters f. w. mean fresh-water, and though sometimes such fishes may descend to the sea, the general habit is meant to be expressed. When two local names for the same fish are in use, the more common one has been chosen, and where the currency for the titles was about the same all have been given. The list would bear considerable extension, especially in the native names, but those only are inserted of the identity of which with defined species I could be quite sure. If the omissions strike readers, it must be remembered that a full list of local names would require a volume to itself.

**A**

Albacore. *Thynnus germo*, a Tunny. Pacific.
Alewife. *Clupea mattawocca*, N. Atlantic.
Allice shad. *Clupea finta*, Europe, R. Nile.
Anchovy. *Engraulis encrassicholus*, Medit.
Angelsea-morris. *Leptochphalus morrisii*, Britain.
Angler. *Lophius piscatorius*. Europ, Am., Africa.
Angel-fish. *Rhina squatina*. N. and S. temp. zone.
Armed bull-head. See Hard-head, second name.
Atherine. *Atherina presbyter* or land smelt, Europe.
Aua, *Agonostoma fosteri*, New Zealand.

**B**

Baccalau. See Klipvisch.
Basking shark. See Pelerin.
Bass. *Labrax lupus*. A sea perch. Europe.
Bastard-dory. In Melbourne. See Boar-fish; also *Enoplosus armatus*.
Bastard trumpeter. In Sydney, *Latris ciliaris*; in Tasmania and Victoria, *L. fosteri*. Also, see Morwong.
Bat-fish. *Psettus argenteus*.
Ballan wrasse. *Labrus maculatus*. Europe.
Banded Morwong. In Sydney, *Chilodactylus vittatus*.
Band-fish. *Cepola rubescens*. Europe.
Barbel. *Barbus vulgatus*. Europe.
Barracouta. *Thyrsites atun*. Southern seas.
Barracudas. *Sphyrænidæ*.
Barramunda. An Australian name for *Osteoglossum leichhardtii*, but also applied to many river fish.
Beardie. In Sydney, *Lotella marginata*.
Becker. *Pagrus orphus*. Britain.
Bellows-fish. *Centriscus scolopax*, Brit. and Tasmania.
Bergylt. *Sebastes norvegicus*, Europe.
Bib. *Gadus luscus*. A cod, N. Atlantic.
Bitterling. *Rhodeina amarus*, a kind of carp. Europe.
Black bass. *Huro nigricans*. A perch of Lake Huron.

FISH AND FISHERIES. 183

Black-fish. In Europe a species of *Centrolophus*; in America a species of Wrasse (*tautoga*); in Sydney, *Girella cuspidata*; in Victoria, *Gadopsis marmoratus*.
Black-head. A species of carp (*Pimephales*). N. America.
Black-horse. Another sp. of carp. (*Sclerognathus*). N. Amer.
Black-bream. In Europe, *Cantharus lineatus*; in Sydney, *Chrysophrys australis*.
Black trevally. In Sydney, *Teuthis nebulosa*.
Bleek and Bleak. *Alburnus alburnus*, a carp. Europe.
Blenny. Applied to all the species of the Blenny family.
Blind-fish. *Amblyopsis spelæus*. Mammoth Cave, Kentucky.
Blue-cod. *Percis colias*, New Zealand.
Blue-fish. *Temnodon saltator*. America. Our "Tailor." In Sydney, *Girella cyanea*.
Blue groper. In Sydney, *Cossyphus gouldii*.
Blue-pointer. In Sydney, *Lamna glauca*, a shark.
Blue shark. *Carcharias glaucus*. Europe.
Boar-fish. In Melbourne, *Histiopterus recurvirostris*; New Zealand, *Cyttus australis*; in Europe, *Capros aper*, one of the Horse-mackerels.
Bogue. *Box boops*. British (*Sparidæ*).
Bolti or Bulti. Species of *Chromis*, from African and Palestine Rivers, *C. andreæ* is the common fish of the River Jordan and the Sea of Galilee.
Bonito. *Thynnus pelamys*, all tropical and temperate seas.
Bony-bream. In Australia, *Chatoessus richardsonii*.
Bony-pike. *Lepidosteus viridis*. America, N. and Cent.
Borer. A lamprey. *Myxine glutinosa*. Europe.
Bow-fin. See Mud-fish.
Bream, sea. In Europe applied to different species of *Sparus*, *Pagrus*, *Pagellus*, *Box*, and *Dentex*, most commonly to *Sparus auratus* of Britain; in Australia, to different species of *Chrysophrys*.
Bream, river. In Europe to a carp, *Abramis brama*; America, *A. americanus*.
Brill. *Rhombus lævis*, Britain; New Zealand, *Pseudorhombus scaphus*.
Brook-trout. *Salmo fontinalis*. America.
Bullahoo. In Sydney, native name for garfish.
Bulls-eye. In Sydney, *Priacanthus macracanthus*.
Bull-head. *Cottus gobio*. Europe.
Bull-head, Sea. *Cottus scorpio*. Europe and America.
Bull-rout. *Centropogon robustus*. Hunter River, N.S.W.
Burbot and Burbolt. *Lota vulgaris*. Europe and America.
Burton-skate, *Raiia marginata*, a ray. Britain.
Butter-fish. *Centronotus gunellus*, a blenny, Britain; in Australia, *Oligorus mitchellii*; in New Zealand, *Coridodax pullus*.
Butterfly-fish. *Gasterochisma melampus*, New Zealand.

C

Capelin. *Mallotus villosus*, Arctic. (*Salmonidæ*.)
Caribe. *Serrasalmo scapularis*. Tropical American rivers. The name is applied to other genera of the family of CHARACINIDÆ, which replaces the Cyprinoids. Small, but very voracious.
Carmoot. A species of *Clarias* (SILURIDÆ). River Nile.

Carp. Applied to many species of *Cyprinus, Carassius,* &c. In Sydney, *Chilodactylus fuscus.*
Cat-fish. Applied to many genera and species of *Siluridæ.* In Europe, *Anarrhicas lupus:* in Sydney, *Cnidoglanis megastoma;* in Australian rivers of the interior, *Copidoglanis tandanus;* in New Zealand, *Kathetostoma monopterygium.*
Chad. Young of Becker.
Charr. *Salmo alpinus,* Europe; in Scotland, *S. killinensis*; in Lake Windermere, *S. willughbii.*
Chub. In Europe, *Leuciscus cephalus.* In the Eastern States of North America about ten species of *Calreichthys* are called chub, likewise *Leucosomus corporalis.*
Climbing-fish. *Periopthalmus australis.*
Climbing perch. *Anabas scandens.* India.
Cloudy Bay cod. *Genypterus blacodes.* New Zealand.
Cod. In Europe and America, *Gadus morrhua*; in Melbourne, *Lotella callarias.*
Coal-fish. *Gadus virens.* North America.
Cobble. In Jersey a local name for Gobies.
Cock and hen paddle. *Cyclopterus lumpus.* North seas.
Cock-up. *Lates calcarifer.* Indian rivers.
Cork-wing. A Wrasse. *Cremilabrus melops.* Britain.
Cook *Labrus mixtus.* Britain.
Coral-fish. *Squamipinnes* and POMACENTRIDÆ.
Craig-fluke. *Pleuronectes cynoglossus,* a flounder. North Seas.
Crooner. *Trigla gurnardus,* Britain.
Crutchet. *Perca fluviatilis.* Europe, f.w.
Crucian carp. *Carassius carassius.* Europe.
Cut-lips. *Exoglossum.* North America.

## D

Dab. A flounder. *Pleuronectes limanda.* Britain.
Dace. *Leuciscus leuciscus,* Europe. *Leucosomus pulchellus,* North America.
Deal-fish. *Trachypterus arcticus.* North Atlantic.
Devil-fish. Rays of the family MYLIOBATIDÆ. In Australia, *Ceratoptera alfredi.*
Dog-fish. Sharks of the family SCYLLIDÆ.
Dolphin. *Coryphæna hippurus.* Northern seas.
Domine. *Epinnula magistratus.* West Indian Barracouta.
Dory. *Zeus faber.* In Sydney, *Z. australis.*
Dragonet. Gobies, of the genus *Callionymus.*
Drum. *Pogonias chromis.* North America (SCIÆNIDÆ.)
Drummer. In Sydney, *Girella elevata.*

## E

Eagle-ray. See Devil-fish.
Eel. In Australia, *Anguilla australis.*
Eel-pout. See Burbot.
Electric eel. *Gymnotus electricus.* Brazil.
Electric ray. Rays, of the family TORPEDINIDÆ.

FISH AND FISHERIES. 185

Electric-sheath or cat-fish. *Malapterurus electricus.* Fresh waters of tropical Africa.
Elephant-fish. *Callorhynchus antarcticus,* New Zealand.
Elleck. *Trigla cuculus.* Britain.
Escholar. *Thyrsites pretiosus.* Havannah.

## F

Fall-fish. See Dace. North America.
Fera. *Coregonus acronius.* Lakes of Europe.
Fighting-fish. *Betta pugnax.* Siam.
File-fish. Species of the genus *Ballistes.* Tropical Seas.
Fishing-frog. *Lophius piscatorius.*
Flat-fish. All species of PLEURONECTIDÆ.
Flat-head. In Australia, all species of *Platycephalus.*
Flat-tail mullet. In Sydney, *Mugil peronii.*
Flounder. In Britain, *Pleuronectes flesus*; in Sydney, *Pseudorhombus russelii*; in Melbourne, *Rhombosolea victoriæ*; New Zealand, *Rhombosolea monopus.*
Fluke. *Rhombus megastoma.*
Flute-mouth. Pipe-fishes. Family SYNGNATHIDÆ.
Flying-fish. Four different species of *Exocoetus.*
Flying-gurnard. Three species of *Dactylopterus.*
Forkbeard. *Merlucius vulgaris.* Europe. Also Hake.
Fortescue. In Sydney, *Pentaroge marmorata,* or 40-skewer.
Frasling. The common Perch. See Crutchet.
Freshwater herring. *Salmo greyi.* Ireland; In Sydney, *Clupea richmondia.*
Freshwater cat-fish. See Cat-fish.
Freshwater sole. *Synaptura selheimi.* Queensland.
Frog-fish. See Fishing-frog; also species of *Antennarius.*
Frost-fish. *Lepidopus caudatus.* Australia, New Zealand. (In Europe this is called Scabbard-fish). Also *Regalecus banksii.*

## G

Gallo. The John Dory is so called in South Europe.
Gar-fish. In Sydney, *Hemirhamphus intermedius*; in Melbourne, *Belone ferox.*
Gar-pike, in America. See Bony-pike. In Britain, *Belone belone*; elsewhere, other species of *Belone.*
Garvie. In Britain, young of *Clupea harengus.*
Gaspereau. See Alewife.
Gillaroo. *Salmo stomachius.* Ireland.
Gilthead. In Britain, *Chrysophrys aurata,* also *Labrus rupestris.*
Globe-fish. Various species of *Diodon* and *Tetradon.*
Glut. *Anguila latirostris.* West Indies.
Glutinous-hag. See Borer.
Goby. Various species of the family GOBIDÆ.
Gold-fish. *Cyprinus auratus.*
Gold sinny. See Cork-wing.
Golden orfe. See Id.
Golden perch. In Australia, *Ctenolates auratus.*

2 A

Gorebill. Britain. *Belone rostrata.*
Gouramy. *Osphromenus olfax.* Indian Archipelago.
Grayling. In Britain, *Thymallus vulgaris;* in Australia, *Prototroctes marœna.*
Great lake trout. *Salmo ferox.* Britain.
Greenback. Britain. *Belone vulgaris.*
Greenbone. See Greenback.
Greenland shark. *Læmargus borealis.*
Grey mullet. *Mugilidæ* generally.
Grey-nurse. A Shark, *Odontapsis americanus.*
Grig. See Glut.
Groper. In Queensland, *Oligorus terræ-reginæ* and *O. goliath*; in New Zealand, see Hapuku.
Growler. *Gristes salmonides* (f.w.) United States.
Guard-fish. Another name for Gar-fish.
Gudgeon. Small fresh-water carps of the genus *Gobio.* In Australia, species of *Eleotris.*
Gunnel-fish. See Butter-fish.
Gurnard. Species of *Trigla.*
Gurnet. See above. In Melbourne, *Centropogon scorpænoides.*
Gwyniad. *Coregonus clupeoides.* Britain.

# H

Haddock. *Gadus æglefinus,* Britain; in New Zealand, *G. australis.*
Hadot. French name for the above, sometimes used in south of England.
Hag-fish. See Borer.
Hair-tail. *Trachiurus lepturus.* Tropics. Like Frost-fish.
Hake. See Fork-beard.
Haku. *Seriola lalandii.* New Zealand.
Half-beak. Species of *Hemirhamphus.*
Halibut. *Hippoglossus vulgaris.* North Seas.
Hammerhead. *Zygæna malleus.* A shark.
Hapuku. *Oligorus gigas.* New Zealand.
Hardhead. *Trigla cuculus.* Britain. Also *Aspidophorus europæus.*
Hardyhead. In Sydney, *Atherina pinguis.*
Hauture. *Trachurus trachurus.* New Zealand.
Hawai. Species of *Eleotris.* New Zealand.
Herring. Various species of *Clupea.*
Hiku. See Frost-fish. New Zealand.
Hipara. See Kokuku.
Hoe-dog. *Acanthias vulgaris.* Europe. Dog-fish.
Hoki. *Coryphænoides novæ zelandiæ.* New Zealand.
Holibut. See Halibut.
Hooknose. See Hard-head, second name.
Homelyn-ray. *Raiia maculata.* Britain.
Hopping-fish. See Climbing-fish.
Horse-mackerel. Species of *Caranx* and *Trachurus*; in Sydney, *Auxis ramsayi.*
Hounds. Small sharks, of genus *Mustelus.*
Hunds-fish. *Umbra krameri,* f.w. Hungary.
Hurling. The common perch. See Frasling.

## I

Id. *Leuciscus idus*. A carp. Central and North Europe.
Ihi. See Gar-fish. New Zealand.
Inanga. Species of *Galaxias*. New Zealand.

## J

Jackass-fish. See Morwong.
Jerusalem-haddock. *Lampris luna*. North Atlantic.
Jew-fish. In Sydney, *Sciæna antarctica*; also *Glacosoma hebraicum*.
John Dory. See Dory.

## K

Kaakaa-lain. See below.
Ka-i-ra. See Murray Cod. Australia.
Kahawai. *Arripis salar*. New Zealand.
Kahawairoa. See Kapuku.
Kanæ. *Mugil perusii*. New Zealand.
Kelp-fish. In New Zealand, *Coridodax pullus*; in Tasmania, *Odax baleatus*.
King-fish. In Britain, *Lampris luna*, also *Chimæra monstrosa*; in Sydney, *Seriola lalandii*; in Melbourne, *Sciæna antarctica*; in Tasmania, *Thyrsites micropus*; West Indies, *Elacate niger*.
King of the herrings. In Britain, *Chimæra monstrosa* and *Regalecus banksii*.
Kiriri. *Macrocanthus convexirostris*. New Zealand.
Kiver. See Sunny.
Klipvisch. *Genypterus* ——. Cape.
Kokopu. *Galaxias* ——. New Zealand.
Kohkohi. See Trumpeter. New Zealand.
Kookoobul. See Murray Cod.
Koobery. In Australia, *Therapon richardsonii*.
Kumu-kumu. *Trigla kumu*. New Zealand.

## L

Lachs. *Salmo salar*. Britain.
Lampern or lamprey. Species of fam. PETROMYZONTIDÆ.
Lancelet. *Branchiostoma lanceolatum*. In Moreton Bay (Queensland), *Epigonichthys pulchellus*.
Lap-lap. A native name for a very small undescribed fish in the fresh-water swamps between South Australia and Victoria.
Launce. *Ammodytes lanceolatus*, Britain; in N.W. Australia, *Conogradus subducens*.
Leather-carp. A variety of *Cyprinus carpio*, without scales.
Leather-jacket. In Sydney, *Monacanthus ayraudi*; in New Zealand, *M. lonvexirostris*.
Lemon-sole. In Britain, *Solea aurantiaca*; in Sydney, *Plagusia unicolor*.
Lesser sand-eel. *Ammodytes tobianus*.
Lesser fork-beard. *Raniceps trifurcas*. Northern seas.
Ling. *Molva vulgaris*. Northern seas; in New Zealand, *Genypterus blacodes*.
Loach. *Misgurnus fossilis*, f. w. Europe; also species of *Nemachilus*, *N. barbatulus* especially.

Loch-leven trout. *Salmo levenensis.* Scotland.
Long-tom. *Belone ferox* in Sydney.
Lump-sucker. *Cyclopterus lumpus.*
Lump-fish. See above.

## M

Mackerel. In Britain, *Scomber scomber;* in Australia, *Scomber antarcticus.*
Mackerel-midge. *Motella microphthalma.* Britain.
Mado. In Sydney, *Therapon cuvieri.*
Mahaseer. *Barbus mosal,* f.w. India.
Makawhiti. See Aua.
Maka. Maori for Barracouta.
Mako. Maori for *Lamna glauca.*
Mango. See above.
Mary-sole. See Fluke.
Maskinonge. In North America, *Exox estor.*
Marare. See Butter-fish. New Zealand.
Matiwhita, a red Kokopu. New Zealand.
Mattie. Scotch name for Herring caught in May and June.
Meagre or maigre. *Sciæna, aquila* temperate seas.
Miller's thumb. See Bull-head.
Minnow. *Leuciscus phoxinus,* Britain; also applied to various kinds of Carp in Europe and America. *Galaxias attenuatus,* New Zealand.
Mishcup, *Pagrus argyrops.* North America.
Moki. *Latris ciliaris.* New Zealand.
Monk-fish. See Angel-fish.
Moon-eye. *Hyodon tergisus,* f.w. North America.
Morwong. In Sydney. *Chilodactylus macropterus.*
Mossbanker. *Clupea menhaden.* North America.
Mud-fish. *Amia calva,* f.w. North America.
Mullet. Various species of MUGILIDÆ, and formerly some of the TRIGLIDÆ.
Muskellunge. See Maskinonge.
Murray cod. In Australia. *Oligorus macquariensis.*

## N

Nefasch. *Distichodus,* a carp. River Nile.
Nerfling. See Id.
Ngoiro. New Zealand. *Conger vulgaris.*
Nonnat. In France, fry of *Atherina.*

## O

Oar-fish. See Frost-fish.
Oil sardine. *Clupea scombrina.* East Indies.
Old-wife. *Cantharus lineatus,* in Britain. In Sydney, *Enoplosus armatus.*
Ombre. *Umbra cirrhosa.* Mediter.
Ombre-chevalier. *Salmo ombla.* Swiss Lakes.
Opah. See Jerusalem Haddock.
Orfe. See Id.
Oulachan. *Thaleichthys.* North Pacific.

## P

Pakirikiri. See Blue-cod. New Zealand.
Pall. See Lancelet.
Paradise-fish. *Polyacanthus viridi-auratus.* Indian Archipelago.
Parr. Young salmon.
Parrot-fish. Various species of *Scarus.* North Atlantic. In Australia, *Pseudoscarus.*
Parrot-wrasse. Same as above.
Parrot-perch. In Australian tropics. *Pseudoscarus.*
Patiki. *Rhombosolea monopus.* New Zealand.
Peal. *Salmo cambricus,* and young of *S. salar.*
Pelerin. *Selache maximus.* Northern seas.
Perch. In Britain. *Perca fluviatilis;* in New South Wales, *Lates colonorum;* in Victoria, species of *Ctenolates.*
Pesce rey. *Atherinichthys laticlavia.* Chili.
Pickerell. In North America, species of Pike, *Esox.*
Pig-fish. In Sydney. *Cossyphus unimaculatus.*
Piharau. Lampreys. New Zealand.
Pike. In Britain, *Esox lucius,* which is the name for the American species. In Sydney, *Sphyræna obtusata* and *Lannioperca.*
Pike-perch. *Lucioperca zandra,* f.w., Europe.
Pilchard. *Clupea pilchardus,* Britain. *Clupea sagax.* New Zealand.
Pilot-fish. *Naucrates ductor.*
Pipe-fish. Species of *Syngnathus.*
Piper. *Trigla lyra.* Britain.
Plaice. *Pleuronectes platessa.*
Pla-kat. See Fighting-fish.
Pocket-fish. See Angler.
Pogge. See Hard-head.
Pohaiakaroa, *Sebastes percoides.* N. Zealand.
Pollack. *Gadus pollachius.* N. Seas.
Pollan. *Coregonus pollan,* f.w. Ireland.
Polwig. Fry of Goby.
Pope. *Acerina cernua,* f.w. N. Europe.
Porbeagle. *Lamna cornubica,* a shark. N. Seas.
Porgy. See Mishcup.
Port Jackson shark, species of *Heterodontus.*
Pout. See Bib.
Powan or powen. See Gwyniad.
Pride. A Lamprey. *Petromyzon branchialus.*
Prussian carp. See Crucian carp.
Pundy. See Murray-cod.
Pumpkin-seed. See Sunny.

## R

Rabbit-fish. See King of the herrings.
Ray. The name applied to the Division BATOIDIE, of the Order CHONDROPTERYGII.
Red-dace. *Leuciscus leuciscus.* Britain.
Red-eye. *Leuciscus erythrophthalmus.* Europe.
Red-fin. See above, also *L. cornutus.* N. America, f.w.
Red-flathead. *Platycephalus bassensis.* Australia.
Red-gurnard. See *Kumu-kumu.* In Melbourne, *Sebastes percoides* also Red Rock-gurnard.

Red-horse. Carp of the genus *Catostoma*. N. America.
Red mullet. Several genera of fam. MULLIDÆ. In N. Zealand *Upenoides vlamingii*.
Red rock-cod. In Sydney, *Scorpæna cruenta* and *S. cardinalis*; N. Zealand, *Percis colias*.
Red-snapper. *Scorpis hectori*. N. Zealand.
Red sole. See Solenette.
Remora. *Echineis remora* or Sucking-fish.
Ribbon-fish. Species of *Trachypterus*. *Regalecus* in temperate and tropical seas.
Rig. See Tope.
Roach. *Leuciscus rutilus*.
Rock-cod. In Melbourne, *Pseudophycis barbatus*.
Rock-cook. *Centrolabrus*. Britain.
Rock-flathead. In Australia. *Platycephalus lævigatus*.
Rock-ling. In Melbourne, *Genypterus australis*.
Rockling. Various species of *Motella*. Temperate seas.
Rock-perch. In Melbourne, *Glyphidodon victoriæ*.
Rock-whiting. In Sydney, *Odax semifasciatus* and *O. richardsonii*.
Rough-dab. *Hippoglossoides*. Britain.
Rough-hound. *Mustelus vulgaris*. Britain.
Rudd. See Red-eye, also male *Salmo salar*.
Ruff. See Pope.

## S

Sail-fish. *Carpioides*. N. America.
Sail-fluke. See Fluke.
Salmon. In N. Hemisphere. Various species of *Salmo*. In Australia, *Arripis salar*.
Salmon-trout. *Salmo trutta*. N. Hemisphere.
Salvelini. The Charr tribe of *Salmones*.
Samson-fish. In Sydney, *Seriola hippos*. In Melbourne, young of *Arripis salar*.
Sand-eel. See Launce.
Sand-smelt. See Atherine.
Sand-piper. See Pride.
Sand-pride. See above.
Sandra. See Pike-perch.
Sandy-ray. *Raiia circularis*. Britain.
Sapphirine-gurnard. *Trigla hirundo*. Britain.
Sar, also Sargus and Saragu, species of *Sargus*. N. Atlantic and Medit.
Sardine. See Pilchard.
Saury-pike. *Scombresox saurus*. Britain.
Saw-fish. Species of Shark. *Pristis*.
Scabbard-fish. See Frost-fish.
Scad. See Horse-mackerel.
Scald-fish. *Arnoglossus lanterna*. Britain.
Schal. Species of *Synodontis*. R. Nile.
Schell-fish. See Haddock.
Schelly. See Powan.
Schnapper. In Australia, *Pagrus unicolor*.
School-shark. In Sydney, *Galeus australis*.
Scissors. Genus of Cypriniod-fish. S. America.

Scorpion-fish. Species of the family SCORPENIDÆ.
Scup. See Misheup.
Sea-bat. Species of *Platax*. Indian Ocean and W. Pacific.
Sea-bream. Species of SPARIDÆ family.
Sea-cat. See Wolf-fish.
Sea-devil. See Devil-fish, also Fishing-frog.
Sea-fox. *Alopecias vulpes*. World-wide.
Sea-gudgeon. See Goby.
Sea-hedgehog. Species of *Diodon*.
Sea-horse. Species of *Hippocampus*.
Sea-needle. See Gar-fish. *Belone*.
Sea-pipe. See above.
Sea-perch. Species of *Serranus*. In Sydney. See Morwong and Bull's eye. *Sebastes percoides*, N. Zealand.
Sea-serpent. See Ribbon-fish.
Sea-sheep. See Maigre.
Sea-snipe. See Bellows-fish.
Sea-trout. See Salmon-trout.
Sea-wolf. See Wolf-fish.
Seeingo. *Percalabrax japonicus*. China seas.
Sergeant-baker. In Sydney, *Aulopus purpurissatus*.
Sewin. *Salmo cambricus*. N. Europe.
Shad. See Allice-shad.
Shagreen-skate. *Raiia fullonica*. Britain.
Shanny. *Blennius photis*. Britain.
Shark. Various species of order CHODROPTERYGII, tribe *Selachoidei*.
Sheat-fish. *Silurus glanis*. Europe.
Sheep's-head. *Sargus ovis*. N. Atlantic.
Silver-bream. In Australia, *Gerres ovatus*.
Silver-eel. In Sydney, *Murænesox cinereus*.
Silver-fish. Variety of Gold-fish.
Silvery-hairtail. See Scabbard-fish.
Silver-perch. *Therapon richardsonii*.
Skate. *Raiia batis*. Britain.
Skipper. See Saury pike.
Skip-jack. *Temnodon saltator*.
Skulpin. *Callionymus draco*. Britain.
Smear-dab. *Pleuronectes microcephalus*.
Smelt. *Osmerus eperlanus*, Northern seas. In Melbourne, *Clupea vittata*.
Smig. A name for Whitebait, which see.
Smooth-hound. *Mustelus lævis*. Britain.
Snake-pipe-fish. *Syngnathus ophidian*. Britain.
Snapper. See Schnapper.
Snig-eel. *Anguila mediorostris*. S. England.
Snoek. See Barracouta.
Sole. In Britain, *Solea vulgaris*; in Sydney, *Synaptura nigra*; in Melbourne, *Rhombosolea bassensis*; in N. Zealand, *Rhombosolea*.
Solenette. *Solea parva*. S. Britain.
Sparling. See Smelt.
Spawn-eater. *Leuciscus hudsonius*. N. America; also called Smelt.
Spear-fish. See Sail-fish.
Sprat. *Clupea sprattus*. Atlantic coasts.

St. Mary's trout. *Motella vulgaris.* Britain.
St. Peter's fish. See John Dorey.
Stare-gazer. Species of *Uranoscopus.*
Starry-ray. *Raiia radiata.* Britain.
Sterlet. *Acipenser ruthenus.* Rivers of Russia.
Stickleback. Species of *Gastrosteus,* f.w. N. Hemisphere.
Sting-bull. *Trachinus draco.* Britain.
Sting-ray and Stingaree. Species of family *Trygonidæ.*
Stock-fish. See Cod, also Hake.
Stone-bass. *Polyprion cernium.* European coasts.
Stone-biter. See Burbolt.
Stone-clagger. See Lump-sucker.
Stone-lifter. In Melbourne, *Kathetostoma læve.*
Stone-lugger. Species of Carp, *Campostoma.* Europe.
Stone-loter. Species of Carp, *Exoglossum.* N. America.
Stone-roller. See Red-horse.
Stranger. In Melbourne, *Odax richardsonii.*
Sturgeon. Species of *Acipenser.* Europe.
Sucker. See Lump-sucker and Remora; also, species of carp in N. America, *Catostomus.*
Sun-fish. Species of PERCIDÆ in N. American rivers; also, *Lampris luna*; also, *Orthagoriscus mola.*
Sunny, *Pomotis vulgaris.* N. America, f.w.
Surgeon. Species of *Acanthurus.*
Sweep. In Sydney, *Scorpis æquipennis.*
Sword-fish. Species of *Histiophorus* and *Xiphias*

## T

Tadpole-fish. See Lesser forkbeard.
Tailor. In Sydney, *Temnodon saltator.*
Tamure. See Snapper. New Zealand.
Tarwhine. In Sydney, *Chrysophrys sarba.*
Tara kahi. See Morwong. New Zealand.
Tasmanian-trout. *Galaxias truttaceus.* Tasmania.
Tautoga or tautog. See Black-fish. America.
Tawa-tawa. New Zealand. Mackerel.
Telescope-fish. Monstrosity of Crucian carp.
Tench. *Tinca tinca.* Europe, f.w.
Thorn-back. See Sting-ray.
Thresher. See Sea-fox.
Tiger-shark. *Stegostoma tigrinum.* Ind. Ocean.
Toad-fish. Species of *Tetrodon.*
Tope. Species of *Galeus,* sharks.
Topknot. *Rhombos punctatus.* Britain.
Torgoch. *Salmo perisii.* Wales.
Torpedo. See Electric ray.
Torsk. *Brosmius brosme.* Northern seas.
Trout. *Salmo fario.* Europe.
Trubu. *Clupea toli.* Indian Archipelago.
Trumpet-fish. See Bellows-fish.
Trumpeter. In Tasmania, *Latris hecateia.*
Trumpeter-whiting. In Sydney, *Sillago bassensis.*

Tub-fish. *Trigla hirundo.* Britain.
Tunny. *Thynnus thynnus.* World-wide.
Turbot. *Rhombus maximus.* N. Atlantic.
Tusk. In Scotland. See *Torsk* and Forkbeard.
Twaite-shad. *Clupea finta.* Europe and R. Nile.

## U

Umbra. See mud-fish.
Umber. See Grayling.
Unctuous sucker. *Cyclopterus montagru.* Britain.
Upokororo. *Prototroctes oxyrhynchus.* N. Zealand.

## V

Vaagmaer. See Deal-fish.
Vendace. *Coregonus vandesius.* Britain.
Viviparous blenny. *Zoarces viviparus.* Europe.
Viviparous dog-fish. *Mustelus vulgaris.* Britain.

## W

Wairepo and Waingengo. New Zealand. Stingaree (*Trygon*).
Warehou. *Neptonemus brama.* New Zealand.
Weever. See Sting-bull, also *Trichinus vipera.*
Wels. *Silurus glanis.* Britain.
Whapuku. See Hapuku.
Whiff. See Fluke.
Whitebait. Fry of herring.
White-fish. Species of *Leuciscus* and *Corregonus.*
White pointer. In Sydney, *Carcharodon rondeletii.*
White trevally. In Sydney, *Caranx georgianus*; also, *Teuthis javus.*
Whiting. In Britain, *Gadus merlangus*; in Sydney, *Sillago maculata* and *S. ciliata;* in Melbourne, *S. punctata.*
Whiting-pout. See Bib.
Wolf-fish. *Anarrhichas lupus.* N. seas.
Wrasses, fishes of family LABRIDÆ.

## Y

Yarra herring. In Australia, *Prototroctes marœna.*
Yarra trout. See Tasmanian trout.
Yawlings. See Whitebait.
Yellow-belly. *Ctenolates auratus.* Australia, f.w.
Yellow-tail. See Horse-mackerel. In Melbourne, *Seriola grandis.*

## Z

Zebra shark. See Tiger-shark.
Zarthe. *Abramus vimba.* Europe.
Zope. *Abramus ballorus.* Europe.
Zuzuki. See Seengo.

2 B

# CHAPTER XVI.
## Works relating to Fish and Fisheries.

The following list is not meant to be a complete catalogue of all the works relating to the subject, for that would be more bulky than the rest of the volume. No more is attempted than to give the student the names of those books which should be consulted to obtain a mastery of the subject, or those which contain useful or interesting information about Fish and Fisheries. A few of the very ancient authors are given, for they may fall within the reach of students, and they are worth consulting if only for the sake of the figures by which they are illustrated. Sometimes the figures are rude and even grotesque, but often they are master-pieces, and at the worst can be generally recognized. Finally, the Australian portion of the list has been made as complete as possible.

Aristotle, 384 to 322 B.C. History of Animals. A Latin edition, with figures, was published in Venice in 1513, folio.

Marcus Terentius Varro; born B.C. 126, died B.C. 26. Said to have written 500 volumes, but two only extant:—De Re Rustica and Analogia, fragments only. In the former there is much information about fishes and fish preserves, worth reading even at the present day.

Lucius Junius Columella; De Re Rustica (on Agriculture). This is about the most detailed work we have on the Roman Fish preserves, from which much may still be learned.

Plinii Historia naturalis, lib. ix, de aquatilium natura. Pliny's Nat. History, 6 vols. Bostock and Riley's translation; Bohn's edit., vol. II. The author died A.D. 79, and the 1st edit. of his work was printed in 1469. In the works large numbers of Roman authors are quoted at p. 237, only a few of which have come down to us.

Oppian's Halieutics. (In Greek.)

Bellonius de aquatilibus. Paris, 1553, folio.

Rondeletius, de piscibus. Lugd., 1554, folio.

Salviani aquatilium animalium historiae. Rom., 1554, folio.

Conr. Gesneri historia piscium, in historia animalium. 1551, folio.

Steph. a Schonvelde Ichthyologia et nomenclatura animalium marinorum, &c. Hamb., 1624, 4to.

Aldrovandus, de piscibus. Bonon. 1638, folio.

Jonstonus, de piscibus.

Franc. Willughby, de historia piscium. Oxf., 1686, folio.

Jo. Raii, Synopsis methodica piscium. Lond., 1713, 8vo.

Petr. Artedi; Ichthyologia, curâ Carl. Linnaei. Lugd. Bat., 1738, 8vo. Curâ Walbaum; Gryp., 1788, 8vo. Curâ Schneider; Leips, 1789, 4to.

Jac. Theod. Klein; historia piscium. Gedan., 1742, 4to.

Marc. Catesby; piscium et serpentum imagines ad naturam expressae. London, 1731, folio.

Laur. Theod. Gronovii; Museum ichthyologicum. Lugd. Bat., 1745, in 2 vols., folio.

Jac. Christ. Schaeffer ; epistola de studii ichthyologici faciliori et tutiori methodo. Ratisb., 1760, 4to.
Ant. Gouan ; Historia piscium. Strasb., 1770, 4to.
Aug. Broussonet ; Ichthyologia, sistens piscium descriptiones et icones. Lond., 1782, 4to.
Jacobus Petiver ; aquatilium Amboinae icones. Lond., 1713, folios
Louis Renard ; poissons, &c., que l'on trouve autour des isle. Moluques, et sur les côtes des terres Australes. Amst., 1764, folio. Very useful in the study of N. Australian fishes.
Schaeffer ; piscium Bavarico Ratisbonensium pentas. Ratisb., 1761, 4to.
Jo. Christ. Wulff ; Ichthyologia regni Borussici. Regiom, 1765, 8vo.
Mait. T. Brunnichii Ichthyologia Massiliensis. Hafn., 1768, 8vo.
The Complete Angler ; by Isaac Walton ; with Notes by Sir John Hawkins. Lond., 8vo., 1802.
Brookes : The Art of Angling, rock and sea fishing, with the natural history of river, pond, and sea fish. 12mo., Lond., 1743.
Pennant's British Zoology, vol. iii.
E. Donovan : The Natural History of British Fishes, in 5 vols. 8vo., Lond., 1808.
Saml. Dale's Natural History of the sea coast and country about Harwich. 4vo., Lond., 1730.
Petiver : Memoirs for the curious, 1708. An account of the English fresh-water fishes.
Borlase's Natural History of Cornwall. Oxf., 1758, folio.
Eleazar Albin : icones piscium. Lond., 1735, 4to.
Carolus a Meidinger icones piscium Austriac indigenorum. Vienn., 1785, folio.
Marc. El. Bloch : Oekonomische Naturgeschichte der Fische Deutschlands. Berlin, 1782, 4to.
Marc. El. Bloch : Naturgeschichte der Auslaendischen Fische. Berlin, 1785, 4to.
Marc. El. Block : Ichthyologie. Berlin, 1785, folio.
Giueseppe Olivi : Zoologia Adriatica, ossia catalogo degli animali del golfo e delle lagune di Venezia. Bassano, 1792, 4to.
Forskahl : Descriptiones animalium, quae in itinere Orientali observavit. Haun., 1775, 4to.
N. G. Leske : Ichthyologia Lipsiensis, 8vo. Lips, 1774.
Car Linnaei : Systema Naturae ; Gmelius edit, translated by Turton, 7 vols. London, Lackington, 1806, only 72 genera of fishes enumerated.
——— Museum ; Adolphi Frideric. Holm., 1754, folio.
Hans Stroem : Physisks og oeconomisk, &c. Son., 1792, in 2 vols., 4to.
Jo. Gottf. Richtes : Ichthyo-theologie, oder Versuch die Menschen, aus Betrachtung der Fische zur Bervunderung ihres Schoepfers aufuhren. Leipz., 1754, 8vo.
La Cepede : L'Histoire naturelle des poissons, in 2 vols. 4to, Paris, 1798.
John Fr. Gronovius : A method of preparing specimens of fish, by drying their skins. Phil. Trans., vol. xcii, p. 57 and 58.

Reaumur : Divers means for preserving dead birds, reptiles, fishes, and insects. Loc. cit. No. 487, p. 304.
The Zoology of New Holland; by I. Ed. Smith. 4to. London, 1798.
Ancient and Modern Fish Tattle; by the Rev. Dr. C. D. Badham. London, Parker, 1854.
Stewart's Natural History. 2 vols. Edinburgh, 1817.
A Review of the Domestic Fisheries of Great Britain and Ireland; by Robert Fraser, Esq. Edinburgh, 1818.
A Treatise on Food and Diet; by Jonathan Pereira, M.D., &c. London, Longman & Co., 1843.
A Treatise on the Management of Fresh-water Fish; by Gottlieb Boccius. London, Van Voorst, 1841.
An Account of Three New Specimens of British Fishes; by Richard Parnell. Royal Society, Edinburgh, 1837.
Angler's and Tourist's Guide; by Andrew Young, Invershire. Edinburgh, A. & C. Black, 1857.
British Fish and Fisheries. Religious Tract Society.
Ceylon, Notes on; by James Stewart, Esq., of Colpetty. Printed for private circulation, 1862.
Couche's Fishes of the British Islands. Groombridge, 1865.
Cuvier's Animal Kingdom; Griffith's edition, Vol. IV, Fishes. London, Whittaker, 1834.
Pisciculture et Culture des Eaux; par P. Trigneaux. Paris, Libraire Agricole de la Maison Rustique.
Pisciculture Pratique et la Multiplication des Sangsues; par Quenard. Paris, De Dusacq, 1855.
Propagation of Oysters; by M. Koste and Dr. Kemmerer. Brighton, Pearce, 1864.
Report by the Commissioners for the British Fisheries of their Proceedings in the Year ended 31st December, 1862, being the Fishing of 1862, and subsequent years.
Reports of the Commissioners of Crown Lands of Canada, 1863–64-65.
Report of the Royal Commissioners on the Operation of Acts relating to Trawling for Herring on the Coasts of Scotland. Presented to both Houses of Parliament by command of Her Majesty, 1863.
Salmon and other Fish, Propagation of; by Edward and Thomas Ashworth. Stockport, E. H. King, 1853.
Seaside and Aquarium; by John Harper. Edinburgh, Nimmo, 1858.
Seaside Divinity; by the Rev. Robert W. Fraser, M.A. J. Hogg & Sons, 1861.
Sketches of the Natural History of Ceylon; by Sir J. Emerson Tennent. London, Longman & Co., 1861.
The Herring; its Natural History and National Importance; by John Metchel, F.R.S., &c. Edinburgh, 1864.
The Young Angler's Guide, &c. London, J. Cheeto, 1839.
Vacation Tourists, 1862–63. London, Macmillan, 1864.
Log Book of a Fisherman; by F. Buckland. London, Chapman & Hall, 1875.

Forrester's Fish and Fishing in the United States. New York, Townsend, 1864.
Guide du Pisciculture ; par J. Remy. Paris, Lacroix, 1854.
Guide Pratique du Pisciculture ; par Pierre Carbonnier. Paris, Lacroix, 1864.
Herring-fishery, on the Existing State of the. '*Herald*' Office, Aberdeen, 1854.
Howitt's Angler's Manual. Liverpool, 1808.
Ichthyonomy. Swinnerton & Brown, Macclesfield, 1857.
L'Alienation des Rivages ; par M. Coste. Paris, 1863.
La Pêche en Eau Douce et en Eau Salée ; par Alphonse Karr. Paris, Michel Levy Frères, 1860.
Multiplication Artificielle des Poissons ; par J. P. J. Koltz. Paris, Lacroix.
Fresh-water Fishes of Central Europe ; by L. Agassiz.
Elements of Nat. Hist. ; by G. Stewart. Edin., 1817. 2 vols. Vol. i., pp. 300 to 430.
Swainson's Fishes, &c. 2 vols., Cab. Ency. London, 1838.
Van der Hoeven's Handbook of Zoology. 2 vols. London, 1858. Vol. ii., pp. 1 to 205.
Fisheries : Article in Encyclop. Brit. 9th edition.
The Harvest of the Sea ; by J. G. Bertram. London, Murray, 1873.
Directions for Taking and Curing Herrings, and for Curing Cod, Ling, Tusk, and Hake ; by Sir Thomas Lauder, Bart. Edinburgh, 1846.
Elements de Pisciculture ; par M. Isidore L'Amy. Paris, 1855.
Experimental Observations on the Development and Growth of Salmon Fry, &c. ; by John Shaw. Edinburgh, A. & C. Black, 1840.
Fish and Fishing in the Lone Glens of Scotland ; by Dr. Knox. Routledge & Co., 1854.
Fish Hatching ; by Frank T. Buckland. Tinsley Brothers, 1863.
Fisheries, The, considered as a National Resource, &c. Dublin, Milleken, 1856.
Ichthyology : Article in Encyclopædia Britannica ; 8th edition ; by Sir John Richardson.
Natural History and Habits of the Salmon, &c. ; by Andrew Young. Longman & Co., 1854.
Natural History of the Salmon, as ascertained at Stormontfield ; by William Brown. Glasgow, Thomas Murray, 1862.
Naturalist's Library ; by Sir William Jardine. Edinburgh, 1843.
Notice Historique sur l'Artificelles de Pisciculture de Huningue. Strasbourg, Berger Levrault, 1862.
Les Huitrières Artificelles de Terrains Emergents ; par M. Coste. Paris.
The Sea and its Living Wonders ; by R. G. Hartwig. London, Longmans.
Observations on the Fisheries of the West Coast of Ireland, &c. ; by Thomas Edward Symons. London, Chapman & Hall, 1856.
Oyster, The : Where, how, and when to find, breed, cook, and eat it. Trübner & Co.

Pisciculture. Pisciculteurs et Poissons; by Eugène Voel. Paris, F. Chamerat, 1856.
Pisciculture et la Production des Sangsues; par Auguste Jourdier. Paris, 1856.
Voyage d'Exploration sur la Littoral de la France et de l'Italie; par M. Coste. Paris, 1861. Imprimerie Impériale.
The Study of Fishes; by Dr. A. Günther. Edinburgh, A. & C. Black, 1880.
British Fishes; by Frank Buckland. London, 1881.
The Commercial Products of the Sea; by P. L. Simmonds. London, 1879.
The Practical Fisherman; by J. H. Keene. London, 1881.
A Book on Angling; by B. F. Francis. London, Longmans; 5th edit., 1883.
The Sea Fisherman. London, Longmans; 3rd edit., 1883.
Blaine's Encyclop. of Rural Sports. London, Longmans.

The following list of works on local Faunas is, with the exception of the Australian list, taken from Günther's work :—

### 1.—GENERAL.

#### French Works.

1. Voyage autour du monde, sur les Corvettes de S.M. l'Uranie et la Physicienne, sous le commandment de M. Freycinet. Zoologie: Poissons; par Quoy et Gaimard. (Paris, 1824, 4to, atlas fol.)
2. Voyage de la Coquille: Zoologie; par Lesson. (Paris, 1826-30, 4to. atlas fol.)
3. Voyage de l'Astrolabe, sous le commandement de M. J. Dumont d'Urville. Poissons; par Quoy et Gaimard. (Paris, 1834, 8vo. atlas fol.)
4. Voyage au Pole Sud; par M. J. Dumont d' Urville. Poissons; par Hombron et Jacquinot. (Paris, 1853-4, 8vo. atlas fol.)

#### English.

1. Voyage of H.M.S "Surphur." Fishes; by J. Richardson. (London, 1844-5, 4to.)
2. Voyage of H.M.S. "Erebus" and "Terror." Fishes; by J. Richardson. (London, 1846, 4to.)
3. Voyage of H.M.S. "Beagle." Fishes; by L. Jenyns. (London, 1842, 4to.)
4. Voyage of H.M.S. "Challenger." Fishes; by A. Günther (in course of publication).

#### German.

1. Reise der österreichischen Fregatte Novara. Fische; von R. Kner. (Wien, 1865, 4to.)

### II.—FAUNÆ.

#### Great Britain.

1. R. Parnell: The Natural History of the Fishes of the Firth of Forth. (Edinb., 1838, 8vo.)

2. W. Yarrll: A History of British Fishes. (3rd edit. Lond., 1859, 8vo.)
3. J. Couch: A History of the Fishes of the British Islands. (Lond., 1862-5, 8vo.)

### DENMARK AND SCANDINAVIA.

1. H. Kroyer: Danmark's Fiske. (Kjobnh, 1838-53, 8vo.)
2. S. Nilsson: Skandinavisk Fauna. (Vol. IV, Fiskarna. Lund., 1855, 8vo.)
3. Fries och Eskträm: Skandinavisk Fauna. (Stockh., 1836, 4to, with excellent plates.)

### RUSSIA.

1. Nordmann: Ichthyologie Pontique, in Voyage dans la Russie méridionale de Demidoff. (Tom. III. Paris, 1840, 8vo, atlas fol.)

### GERMANY.

1. Heckel and Kner: Die Susswasser-fische; von Mitteleuropa. (Leipz., 1863, 8vo.)

### ITALY AND MEDITERRANEAN.

1. Bonaparte: Iconografia della Fauna Italica. (Tom. III. Pesci. Roma, 1832-41, fol., incomplete.)
2. Costa: Fauna del Regno di Napoli. Pesci. (Napoli, 4to, about 1850, incomplete.)

### FRANCE.

1. E. Blanchard: Les Poissons des eaux douces de la France. (Paris, 1866, 8vo.)

### PYRENEAN PENINSULA.

The fresh-water Fish-fauna of Spain and Portugal was almost unknown, until F. Steindacher paid some visits to those countries for the purpose of exploring the principal rivers. His discoveries are described in several papers in the Sitzungsberichte der Akademie Zu Wien. B. du Bocage and F. Capello contributed towards the knowledge of the marine fishes on the coast of Portugal. (Journ. Science Acad. Lisb.)

### NORTH AMERICA.

1. J. Richardson: Fauna Boreali Americana. (Part III. Fishes. Lond., 1836, 4to.) The species described in this work are nearly all from the British possessions in the North.
2. Dekay: Zoology of New York. (Part IV. Fishes. New York, 1842, 4to.)
3. Reports of the United States Commission of Fish and Fisheries. (5 vols. Washington, 1873-79, 8vo. In progress. Contains most valuable information.)

Besides these works, numerous descriptions of North American fresh-water fishes have been published in the reports of the various U. S. Government expeditions, and in North American scientific journals, by Storer, Baird, Girard, W. O. Ayres, Cape, Jordan, Brown, Goode, &c.; but a good general, and especially critical, account of the fishes of the United States is still a desideratum.

JAPAN.

1. Fauna Japonica. Poissons; par H. Schlegel. (Ludg. Bat., 1850, fol.)

.EAST INDIES : TROPICAL PARTS OF THE INDIAN AND PACIFIC OCEANS.

1. E. Ruppell Neue Wirbelthiere. Fische. (Frankf., 1837, fol.)

"These two works form the standard works for the student of the Fishes of the Red Sea, and are distinguished by a rare conscientiousness and faithfulness of the description and figures ; so that there is no other part of the tropical seas with the fish of which we are so intimately acquainted as with those of the Red Sea. But these works have a still wider range of usefulness, inasmuch as only a small proportion of the fishes is limited to that area, the majority being distributed over the Indian Ocean into Polynesia. Ruppell's works were supplemented by the first two of the following works :—

3. B. L. Playfair and A. Günther: The Fishes of Zanzibar. (London, 1866, 4to.) ; and
4. C. B. Klunzinger: Synopsis der Fische des Rothen Meers. (Wien, 1870-1, 8vo.)
5. T. Cantor: Catalogue of Malayan Fishes. (Calcutta, 1850, 8vo.)
6. F. Day : The Fishes of India (London, 1875, 4to., in progress); contains an account of the fresh-water and marine species, and is not yet complete.
7. A. Günther : Die Fische der Sudsee. (Hamburg, 4to., from 1873, in progress.)

"Unsurpassed in activity, as regards the exploration of the fish fauna of the East Indian Archipelago, is P. Bleeker, a surgeon in the service of the Dutch East Indian Government (born 1819, died 1878), who from the year 1840, for nearly thirty years, amassed immense collections of the fishes of the various islands, and described them in extremely numerous papers, published chiefly in the Journals of the Batavian Society. When his descriptions and the arrangement of his materials evoked some criticism, it must be remembered that at the time when he commenced his labours, and for many years afterwards, he stood alone, without the aid of a previously named collection on which to base his first researches, and without other works but that of a Cuvier and Valenciennes. He had to create for himself a method of distinguishing species and describing them, and afterwards it would be difficult for him to abandon his original method and the principles by which he had been guided for so many years. His desire of giving a new name to every individual, to every small assemblage of species, wherever practicable, or of changing an old name, detracts not a little from the satisfaction with which his works would be used otherwise.

"It is also surprising that a man with his anatomical knowledge and unusual facilities should have been satisfied with the merely external examination of the specimens ; but none of his numerous articles contain anything relating to the anatomy, physiology, or habits of the fishes which came under his notice ; hence his attempts at systematic arrangement are very far from indicating an advance in Ichthyology.

"Soon after his return to Europe (1860), Bleeker commenced to collect the final results of his labours in a grand work, illustrated by coloured plates—Atlas Ichthylogique des Indes Orientales Néerlandaises (Amsterd., fol., 1862)—the publication of which was interrupted by the author's death in 1878."

### Africa.

1. A. Günther : The Fishes of the Nile; in Petherick's Travels in Central Africa. (London, 1869, 8vo.)
2. W. Peters : Naturwissenschaftliche Reise Nach Mossembique, IV. Flussfische. ˙ (Berlin, 1868, 4to.)

### West Indies and South America.

1. L. Agassiz : Selecta genera et species Piscium, quæ itinere per Brasiliam, collegit J. B. de Spix. (Monach, 1829, fol.)
2. F. de Castelnau : Animaux nouveaux ou rares, recueillis pendant l'expédition dans les parties centrales de l'Amérique du Sud. Poissons. (Paris, 1855, 4to.)
3. A. Günther : An account of the Fishes of the States of Central America. (In Trans. Zool. Soc., 1868.)
4. L. Vaillant and F. Bocourt : Mission scientifique au Mexique et dans l'Amérique centrale. Poissons. (Paris, 1874, 4to. In progress.)

F. Poey, the celebrated naturalist of Havannah, devoted many years of study to the Fishes of Cuba. His papers and memoirs are published partly in two periodicals, issued by himself, under the title of Memorias sobre la Historia natural de la Isla de Cuba (from 1851), and Repertorio Fisico-natural de la Isla de Cuba (from 1865), partly in North American scientific journals. And, finally, F. Steindachner has published many contributions, accompanied by excellent figures, to our knowledge, of the fishes of Central and South America.

### New Zealand.

1. F. W. Hutton and J. Hector: Fishes of New Zealand. (Wellingt., 1872, 8vo.)

### Arctic Regions.

1. C. Lütken : A revised Catalogue of the Fishes of Greenland, in Manual of the Natural History, Geology, and Physics of Greenland. (Lond., 1875, 8vo.)

Although only a nominal list, this catalogue is useful, as it contains references to all the principal works in which Arctic Fishes have been described. The fishes of Spitzbergen were examined by A. J. Malmgren (1865).

### Australia.

Works on Australian fishes are not enumerated by Dr. Günther. The list given in the introduction is far from being exhaustive. They need not be repeated here, but the following additional details will be useful.

The following papers are all from the Proceedings of the Linnean Society of N. S. Wales:—

Notes on the Entozoa of a Sun-fish ; by W. Macleay, F.L.S. ; page 12, vol. I.

Notes on the Zoology of the " Chevert" Expedition ; by W. Macleay, F.L.S. ; page 36, vol. I.

The Ichthyology of the "Chevert" Expedition; by Haynes Gibbes Alleyne, M.D., and William Macleay, F.L.S. (with plates); page 261, vol. I.

The Ichthyology of the "Chevert" Expedition (2); by Haynes Gibbes Alleyne, M.D., and William Macleay, F.L.S. (with eight plates); page 321, vol. I.

Note on Monacanthus Cheverti; by William Macleay, F.L.S.; page 69, vol. II.

Australian Fishes, new or little known; by Count F. de Castelnau; page 225, vol. II.

The Fishes of Port Darwin; by W. Macleay, F.L.S.; plates 7, 8, 9, 10, page 344, vol. II.

Note on a species of Therapon found in a dam at Warialda; by William Macleay, F.L.S; with Remarks by the Rev. J. E. Tenison-Woods, F.G.S., F L.S., &c.; page 15, vol. III.

Descriptions of some new Fishes from Port Jackson and King George's Sound; by William Macleay, F.L.S. (with plates); page 33, vol. III.

Notes on the Fishes of the Norman River; by Count F. de Castelnau; page 41, vol. III.

On some new Australian (chiefly Fresh-water) Fishes; by Count F. de Castelnau; page 140, vol. III.

On a new Ganoid Fish from Queensland; by Count F. de Castelnau; plate 19A, page 164, vol. 3.

On a species of *Amphisile*, from the Palau Islands; by William Macleay, F.L.S.; plate 19B, page 165, vol. III.

—*Plagiostomata* of the Pacific; by N. de Miklouho-Maclay, and William Macleay, F.L.S.; Part I (with five plates), page 306, vol. III.

Essay on the Ichthyology of Port Jackson; by Count F. de Castelnau; page 347, vol. 3.

Notes on some Fishes from the Solomon Islands; by William Macleay, F.L.S.; page 60, vol. IV.

On the *Clupeidæ* of Australia; by W. Macleay, F.L.S., &c.; page 363, vol. IV.

On the *Mugilidæ* of Australia; by William Macleay, F.L.S., &c.; page 410, vol. IV.

Description of a new species of *Galaxias*, with remarks on the distribution of the genus; by William Macleay, F.L.S., &c.; page 45, vol. V.

On two hitherto undescribed Sydney Fishes; by William Macleay, F.L.S., &c.; page 48, vol. V.

Description of a new species of *Oligorus*; by E. P. Ramsay, F.L.S., &c.; plate 9, page 93, vol. V.

Notes on *Galeocerdo rayneri;* by E. P. Ramsay, F.L.S., &c.; plate 4: page 95, vol. V.

On a rare species of Perch from Port Jackson; by E. P. Ramsay, F.L.S., &c.; page 294, vol. V.

Notes on *Histiophorus gladius* (with plate); by E. P. Ramsay, F.L.S.; page 295, vol. V.

Descriptive Catalogue of the Fishes of Australia; by William Macleay, F.L.S., &c.; page 302, vol. V.

Description of two new species of Australian Fishes; by E. P. Ramsay, F.L.S. ; page 462, vol. V.
Description of a Parasitic *Syngnathus*; by E. P. Ramsay, F.L.S. page 494, Vol. V.
Descriptive Catalogue of Australian Fishes ; by William Macleay, F.L.S. ; Part II, plates 13 and 14, page 510, vol. V.
On a new species of *Regalecus*, from Port Jackson ; by E. P. Ramsay, F.L.S. ; plate 20, page 631, vol. V.
Descriptive Catalogue of the Fishes of Australia ; by William Macleay, F.L.S., &c. ; plates 1 and 2, page 1, vol. VI.
Description of a new Labroid Fish of the genus *Novacula*, from Port Jackson ; by E. P. Ramsay, F.L.S., C.M.Z.S., &c. ; page 198, vol. VI.
Descriptive Catalogue of Australian Fishes ; by William Macleay, F.L.S., &c. ; Part IV, page 202, vol. VI.
Description of a new species of *Hemerocates?* from Port Jackson ; by E. P. Ramsay, F.L.S. ; page 575, vol. VI.
On the occurrence of *Pseudophycis breviusculus*, Richardson, in Port Jackson ; by E. P. Ramsay, F.L.S., C.M.Z.S., &c. ; page 717, vol. VI.
On a new species of *Gobiesox*, from Tasmania ; by E. P. Ramsay, F.L.S., &c., vol. VII., page 148.
On the brain of *Galeocerdo rayneri* ; by W. A. Haswell, B.Sc., vol. vii, p. 210, &c.

# INDEX.

N.B.—References are made to the families in a darker type, and where details are given the page numerals are of the same character.
An asterisk distinguishes those fishes figured in the work.

## A

Abdominales, 6.
Acanthias, 25, 93.
Acanthoclinus, 12.
Acanthopterygii, 6, 14, 20, 30, 37, 50, 57, 73.
Acclimatization of fishes, 150.
Acrania, 3.
Acrodus, teeth of, 10.
**Acronuridæ**, 11, 17.
Agassiz, M. Louis, Classification of Fishes, 3.
Agenor modestus, 15.
Air-bladder, 5; in deep-sea fishes, 5; gas of, 5; office of, 5; sometimes has the character of a lung, 5; sometimes wanting, 5.
Alopecias, 25, 92.
Ambassis, 14.
Ammotretis, 12.
Anacanthini, 21, 30, 75.
Anarrhicas lupus, 27.
Anatomy of Fishes, 3–10.
Anchovy, 11.
Anema, 12.
Angel-fish, or shark, 93, 98.
Angler-fish, see Lophius.
Anguilla australis, 88, 107, 139.
Anomalies, apparent in Australian Fish Fauna, 10.
Anthias, 13.
Antennarius, 18, 26.
————— marmoratus, 18.
————— striatus, 18.
————— pinniceps, 18.
————— coccineus, 18.
**Atherinidae**, 20.
* Anthias longimanus, 12, 14, 33.
Apharius roscus, 15.
Aploactis, 11.
————— milesii, 16.
Apodes, 7.

Apogon, 12.
————— fasicatus, 14.
————— guntheri, 14, 107.
————— novae-hollandiae, 107.
Aristeus fluviatilis, 19.
————— lineatus, 19.
Arripis, 12.
————— trutaceus, 36.
————— georgianus, 14.
* ————— salar, 15, 35; poisonous, 36.
Arius, 22.
* Astacopsis, 125.
Atherina, 20.
Atypus, 15.
* Aulopus purpurissatus, 11, 22, 80.
Australian Fishes, number of, 2; exclusively, 12.
Auxis ramsayi, 18.

## B

Barracouta, 56.
Batrachus, 12.
Bat-fish, 18.
**Batrachidæ**, 18.
* Beardy, N.S.W., 21, 74.
Bellows-fish, 11.
Belone, 22, 83.
* ————— ferox, 22, 83.
————— gracilis, 22.
**Berycidæ**, 11, 17, 56.
* Beryx affinis, 17, 50, 51.
. * Black-fish, 11, 21, 105.
**Blenniidæ**, 11, 19, 27.
Blennius, 19, 27, 28.
————— unicornis, 19.
————— castaneus, 19.
Blenny, 27.
Blood of Fishes, 3.
Bonito, 64.
Bony Bream, 23, 106.

Books on Fish and Fisheries, 194–203.
Dorichthys, 13.
────── variegatus, 18.
Brachionichthis, 12.
Brama raii, 18.
Branchial arches, 4.
Branchiostoma, 13, **26**.
Brachypleura, 12.
Bream, sea, 38.
────── silver, **43**.
────── black, **43**.
────── Mr. Hill on, 44.
Brisbania, 85.
Bull's-eye, 35.
Bull-rout, 16, **48**, 107.

## C

Callionymus calauropomus, 12, 19.
────── calcaratus, 19.
────── lateralis, 19.
Canning of fish, 125, 146.
────── at Columbia R., 146.
**Carcharidæ**, 11, 24.
Carcharodon, 25, 92.
**Carangidæ**, 11, 17, 57.
Caranx, 12, 13, 59.
────── nobilis, 17.
────── georgianus, 17, 56, **59**.
────── hippos, 17.
────── ciliaris, 17.
Carassius, 83.
Carcharias, 12, 25, 92, **93**.
* Carp, **46**, 83, 156.
Carpentaria watershed, fishes of, 12.
**Cataphracti**, 11, 19.
Cat-fish, freshwater, 22, **105**.
────── sea, 22, **81**.
Centriscus scolopax, 11.
Centropogon australis, 16.
────── robustus, 16, 47, **108**;
spines of venomous, 48.
Cephalopods, 122.
Ceratodus, a living ganoid, 10.
Cestracion, 10, 96.
────── fossil form of, 96.
**Cestraciontidæ**, 25.
Chanos, 12, 88.
Chætodon strigatus, 15.
────── oliganthus, 15.
Chatoessus, 23, 85, **106**.
Chelmo truncatus, 15.
Chilobranchus, **12**.

\* Chilodactylus fuscus, 11, 13, 16, **46**.
\* ────── macropteris, 16, **47**.
\* ────── vittatus, 16, **47**.
Chiloscyllium, 12.
Chirocentrus, 12, 23.
Chondropterygii, **5**, 24; notochord of, **6**.
Chrysophrys, 12, 16, **42**.
────── sarba, 16.
────── australis, 16.
**Cirrhitidæ**, 11, 16, 45.
Classification—Linnean, **6**.
────── Gunther's, **3**.
────── L. Agassiz's, **3**.
────── Bleeker's, 200.
Clupea, 13, 23, **85**, 86.
────── sagax, 23, **86**.
────── sundaica, 86.
────── novæ-hollandiæ, 23, **87**, 107, 138.
────── sprattus, 11, 85.
────── shoals of, 148.
**Clupeidæ**, 11, 23, **84**.
Cnidoglanis lepturus, 22.
Cod, 7, 21, 75, 89, 156, 168.
────── rock, 14, **33**.
────── Murray, 15, 89, **102**, 107.
────── Melbourne, 21.
\* ────── red rock, 14, 16, **47**.
Conger, 11, 13, 23.
Copidoglanis tandanus, 22, **105**, 106, 107, 138.
Coral fishes, **73**.
Coridodax, 12.
Coris lineolata, 21, 74.
Cormorants, destruction of, **178**.
Corregonus, 14, 109.
Coryphæna punctulata, 18.
**Coryphænidæ**, 11, 62.
Cossiphus, 21, 28, 74, 75.
\* ────── unimaculatus, 21, 74, **75**.
\* ────── gouldii, 21, **74**.
Cottidæ, 11, **67**.
Cox, Dr., on the oyster, 110, 111.
\* Crab, the sea, **125**.
Crangon, 126.
Craptolus, 12.
\* Cray-fish, **125**.
\* ────── freshwater, **126**; eggs of 127.
Crepidogaster, 12.
Cristiceps, 13, 19, **28**.
Crossorhinus barbatus, 25, 93.
Crucian carp, 83, 156.

Crustacea, 125.
Ctenoids, 3.
Ctenolates, 13.
———— ambiguus, 15, 103.
———— christyi, 15, **103**.
———— flavescens, 15.
Cuba, fishes of, 20.
Cuttle-fish, 122, **123**.
Cuvier, 2.
Cybium commersoni, 18.
———— guttatum, 18.
Cycloids, 3.
Cyclostomata, 3, 5, 30; optic nerve of, 4.
**Cyprinidæ**, 83.
**Cyttidæ**, 11, 18, **61**.

### D

Dactylopteres, 19.
Diodon, 12, 21, 91.
Diplocrepis, 12.
Dog-fish, 25, 92.
Dolphin, 7.
\* Dory John, 7, 11, 18, **61**.

### E

Echeneis remora, 18, 64.
Eel family, 87.
—— conger, 23.
—— green, 23, 89.
—— silver, 23, **88**.
Eggs of fishes, 8.
—————— impregnation of, 8, 150.
—————— artificial hatching of, 8, 150, 151.
Elacate nigra, 18, 64.
Eleotris coxii, 19.
———— australis, 107.
———— compressus, 19.
———— grandiceps, 19.
———— mastersii, 19.
Elops, 12.
Encyclopedia Britannica, the, 2.
Engræus, 126.
Engraulis, 11, 13.
Enoplosus, 12.
\* ————— armatus, 14, **32**.
Equatorial fishes in Australia, 10.
————— genera on N.S.W. coast, 12.
Eyes of fishes, 7; influence of light on, 7; eyes of Polynemus, 7.

### F

Families of fishes in N.S.W., 11.
Fiddler, 25, 101.
Fins, value of in classification, 6; correspond to limbs, 6; pectoral, ventral, dorsal, anal, 6; rays and spines of, 6.
Fish, what is a, 2.
Fishing, deep-sea, 143.
———— suggestions as to, 143.
———— line, improvement of, 144.
———— net, 145.
Fishing-grounds of N.S.W., 128-138.
—————————— number and variety of, 128.
"Fish and Fisheries," object of the work, iii; scope of, 2.
Fish fauna of North Coast, 12.
———————— South, 12.
Fishes protected by law, 167.
Fisheries office, Sydney, 149.
**Fistularidæ**, 20.
Flathead, 18, **67**.
———— red, 18, 67.
Flat fishes, 76.
\* Flounder, 21, **76**.
Flying fish, 7.
Food, fishes as, 30.
Fortescue, the, 16, **49**.
Fresh-water fish fauna of N.S.W., 102-109.
Frog-fish, 18, **26**.

### G

**Gadidæ**, 11, 21, **75**.
**Gadopsidæ**, 11, 21.
Gadopsis marmoratus, 21, **105**.
———— an extraordinary fish, 105.
———— gracilis, 105.
———— gibbosus, 105.
Galaxias attenuatus, 14.
———— bong bong, 22.
———— coxii, 22.
———— kreftii, 22.
———— nebulosa, 22.
———— planiceps, 22.
———— punctatus, 22.
———— distribution of, 13.
———— findlayi, 107; Von Muller on, 107.
Galeocerdo, 25, 92.
Galeus australis, 25, 92.

Ganoids, 3.
* Gar-fish, 22, 83.
Genera, north temperate, reappear in N.S.W., 11.
Geriyoroge bengalensis, 14.
Gerres, 12.
—— ovatus, 15, **43**.
—— subfasciatus, 15.
—— argyreus, 15.
Gill-cover. See Operculum.
Gills, peculiar, 5.
—— purpose of, 4.
—— anatomy of, 4.
Girella, 11.
* —— tricuspidata, 16, **39**.
—— simplex, 15.
—— elevata, 16.
—— cyanea, 16.
—— ramsayi, 16.
Glaucosoma scapulare, 14, 34.
—— hebraicum, 54.
**Gobidæ**, 11, 19, 27.
Gobies, 7, **27**.
—— river, 19, 106.
Gobius bifrenatus, 19.
—— buccatus, 19.
—— cristatus, 19.
—— flavidus, 19.
—— semifrenatus, 19.
Golden perch, 15, **103**.
Grapsus, 124.
Grayling, 11, **109**.
Groper, 21, 28.
* —— the blue, 21, **74**.
Gurnard, 7.
—— flying, 19, **68**.
—— Melbourne. See Centropogon.
Günther, Dr., Catalogue of Fishes, 1, 2.
—— Classification of Fishes, 3.
—— statement of, questioned, 13.
—— study of fishes, 5, 20, 31, 198.
Gymnodontes, 11, 24, **28**.

# H

Hake, 7.
Haliotis nævosa, 92, 122.
—— value of, 121.

Haplodactylus, 13.
—— lophodon, 16.
—— obscurus, 16.
Haplochitonidæ, 109.
Hapuku, 102.
Heart of fishes, 3, 4.
—— auricle of in palæichthyes, 4.
Heliastes, 12, 21.
Hemirhamphus regularis, 22, 83.
—— commersoni, 22, 83, 84.
—— argenteus, 22, 83, **84**.
* —— intermedius, 22, 83, 84.
Herring, 7, 23, **85**.
—— of N.S.W., 23, 86.
—— southern, **86**.
—— of W. Australia, **107**.
Heterocercal. See tail.
Heterodontus, 11, 25, 96.
—— dentition of, 10.
Hill, Mr. E. S., **2**, 40, 50, 53, 58, 79.
Hippocampus, 12, 23, 29.
—— novæ hollandæ, 23.
—— peculiarity of, 23.
Histiophorus gladius, 17, **55**.
Holoxenus, 12.
Homocercal. See tail.
**Hoplognathidæ**, 45.
Hypos, 25.
Hypural bone, 6.

# I

India, genera of, common to Australia, 13.
* Inter-operculum, 8, 9.
Intestinal tract, 5.

# J

Jackass-fish, **46**.
Jew-fish, 17, **53**, 55.
—— silver, **55**.
—— Mr. Oliver on, 55.
Jugulares, 7.

# K

Kaira, 106.
Kathetostoma, 12.
Kelp-fish, 21, 75.
* King-fish, 17, **59**.
Kookoobul, 15, **102**.

## L

Labrichthys, 12, 28.
**Labridæ,** 11, 21, 28, **73.**
———— N.S.W. species of, 11.
Labyrinthici, 5.
Lamna glauca, 25, 92, **95.**
**Lamnidæ,** 11, 25.
Lampreys, 3.
* Lanioperca mordax, 12, 20, 69.
* Lates colonorum, 13, 14, **31,** 49, 107.
——— calcarifer, 32.
——— curtus, 14.
——— ciliaris, 16.
——— ramsayi, 14.
Latris ciliaris, 16, 46, **47.**
——— fosteri, 46.
Laws, the fishery, N.S.W., 160.
——— Mr. Oliver on, 160.
——— as to nets, 162.
——— close season, 162.
——— private fisheries, 165.
——— as to weight of fish, 168.
——— oysters, 170.
* Leather-jacket, 24, **89.**
Lemon sole, 22.
Lepidoblennius, 12, 19.
Leptocardii, 3, 5, 26, 30.
Leptoscopus, 12.
Lepidotrigla papilio, 19.
Lethrinus, 12, 16.
* Ling, 76.
Liver of fishes, 5.
Lobotes auctorum, 15.
Long-fin, **33.**
* Long Tom, 22, **83.**
Lophius piscatorius, 11, 26.
Lophobranchii, 23, 30.
Lophonectes, 12.
Lophorhombus, 12.
Lotella, 11, 21, **76.**
* ——— marginata, 21, **76.**

## M

* Mackerel, 7, 18, **62,** 107, 148.
Macleay, Hon. W., 2, 7, 104, 106.
Macquaria, 13, 15.
**Macruridæ,** 11, 76.
Maray, 84.
——— shoals of, 148.
Marine food fishes, 30-101.
Market, Sydney Fish, 139; supplies of, 139, 140; prices in, 140.

M'Coy, Prof., i, 68, 105.
Megalops, 88.
Milt, functions of, 8.
——— position of, 8.
Mitchell R., new species discovered at, 12.
Mixinoids, 7.
Mollusca, 122.
* Monocanthus, 12, 13, 24, **89.**
Monocentris, 17.
* Morwong, 16, **46.**
——— the red, **46.**
Muciferous system, **8.**
Mud-fish of N. Zealand, 108.
* Mugil grandis, 20, 70, **71.**
——— cephalotus, 20, 70.
——— petardi, 107.
——— dobula, 20, 70, 107.
**Mugilidæ,** 11, 20, **69.**
Müller, Baron von, 108.
Mullet, 7, **70.**
——— to capture, 147.
——— red, 38.
* ——— sea, 20, 69, **71.**
——— grey, 20, **69.**
——— sand, 70.
Mullus barbatus, 38.
Murœna afra, 23, **89.**
——— picta, 23.
Murænichthys, 13.
Murænesox, 12.
——— bagoo, 88.
——— cinereus, 23, 88.
**Murænidæ,** 11, 23, 88.
* Murray cod, 13, 15, **102,** 107.
Murrayia güntheri, 13, 15, **104.**
——— cyprinoides, 15, 104.
——— bramoides, 15, 104.
——— riverina, 15, 104.
Murœna, 12.
Mutton-fish, 92, **122.**
Myobatidæ, 26.
Mytilus, 122.
Myxines, 3.
Myxus elongatus, 20, 70, 71.

## N

Naucrates ductor, 18, 64.
**Nandidæ,** 11, 15.
* Nannygai, the, 17, **51,** 52.
Nannocampus, 12.
Nannoperca australis, 14.
——— riverinæ, 14.

Neochanna, 108.
Neoplotosus, 106.
Neptomenus travale, 17.
New South Wales, species peculiar to, 12, 13.
New Zealand, genera common to Australia, 13.
—————— South Australia, 12, 13.
—————— N. S. Wales, 12, 13.
Nipper, 126.
North Temp. Zone, genera of, reappear in N.S.W., 11.
**Notidanidæ**, 25.
Notandanus, 12, 25, 92.

O

Octopus, 122.
Odax, 12, 21, 75.
—— richardsoni, 21, 75.
—— brunneus, 21, 75.
—— baleatus, 21, 75.
Odontaspis, 25, 92, 95.
\* Old wife, 14, 30, **32**.
Olesthcrops, 12.
Oligorus, 13, 15, 89, **102**, 107.
\* ———— macquariensis, 15, 89, **102**, 107.
———— mitchelli, 15, **103**.
———— terræ reginæ, 102.
———— gigas, 102.
Oliver, Mr. A., on the fishery laws, 2, 160.
————————— fishing-grounds &c., 133.
Onchorrhyncus, 153.
\* Operculum, 7, 9.
**Ophidiidæ**, 76.
**Ophiocephalidæ**, 20.
Opisthognatus, 18.
Optic nerves of fishes, 4.
————— palæichthes, 4.
Ostracion, 12, 13, 24.
Otolithus atelodus, 17, 54.
Oyster, **110**.
—— enemies of, 119.
—— anatomy of, 112, 113.
—— Dr. Cox on, 111.
—— the drift, **110**.
—— male and female, to distinguish, 113.
—— ova of, 113, 114.
—— the rock, **111**.
—— Commission for N.S.W., 2.

Oyster culture, 112.
—————— at L. Fusaro, 116.
—————— Kemmerer, "tile," 115.
—————— history of, 115.
—————— at Ile de Ré, 118.
—————— at Auray, 119.
—————— books on, 121.
—————— suggestions as to, 112.
—— fattening, 116.
—————— Mr. C. Pennell on, 117.
—————— Bertram on, 118.

P

Pachymetopon, 16.
\* Pagrus, 13, 16, 39–42.
——— unicolor, 16, **39**.
——.— guttulatus, 40.
Palæichthyes, 3, 4, 24, 30.
Palemon, **126**.
\* Palinurus, **125**.
Parma, 20.
Patæcus, 12.
Patella, 122.
Pediculati, 11, 18, 26.
**Pegasidæ**, 29.
Pegasus, 29.
Pelagic fishes, 10.
—————— not included in present work, 11.
Pelamis australis, 18.
Peltorhamphus, 12.
Pempheris, 12.
————— compressus, 17.
————— macrolepis, 17.
Penæus, 126.
Pentapus, 12.
—————— setosus, 15.
—————— paradiseus
Penturoge marmorata, 12, 16, **49**.
**Percidæ**, 11, **31**, 101, 103, 104.
Perch, **31**.
—— red, 14, 33.
—— golden, 15, **103**.
—— silver, 15, 103.
—— habits of, 32.
—— bait for, 32.
Periophthalmus, 27.
Phyllopteryx, 12, 23, **29**.
—————— an extraordinary fish, 29.
Physostomi, 22, 30, 80.

\* Pig-fish, 21, **75.**
\* Pike, 6, **69.**
Pilot-fish, 18, 64.
Pipe-fish, 7.
Pisciculture, 150.
———— history of, 150.
———— stocking ponds for, 157.
———— books on, 159.
———— Frank Buckland on, 150, 151.
Placoids, 3.
Plagyodus, 28.
\* Platycephalus, 12, 18, **67.**
———— bassensis, 18, **67.**
———— fuscus, 18, **67.**
Platystethus, 12.
Plectognathi, 24, 31.
Plectropoma, 12, **34.**
———— annulatum, 14.
———— cyneostigma, **34.**
———— ocellatum, 14.
———— semicinctum, 14.
———— susuki, 14.
Plesiops, 15.
**Pleuronectidæ,** 11, 21, **76,** 79.
Plotosus, 106.
Polynemus, 52.
———— indicus, 52.
———— macrochir, 52.
———— cæcus, 7.
**Polynemidæ,** 11.
**Pomacentridæ,** 11, 20.
Pomacentrus, 12, 20.
Porcupine, 91, 89.
\* Pre-opeculum, 8, 9.
Prawn, 126.
Priacanthus macmeanthus, 14.
———— benmembari, 14.
**Pristiophoridæ,** 11, 25.
Pristiophorus, 11, 25, 93, **98.**
Prionurus, 17.
Prototroctes, 14, **109.**
Psenes leucura, 18.
Psettus argenteus, 12, 18.
Pseudoambassis castelnaui, 14.
———— ramsayi, 14.
———— jacksoniensis, 14.
Pseudophycis, 12.
\* Pseudorhombus russellii, 21, 77.
———— multimaculatus, 21, 77.
Pterois volitans, 16.
———— zebra, 16.
Pyloric appendages, 5.
Pundy, 15, 102.

## R

Raiia, 11, 13.
**Raiidæ,** 11, **99,** 101.
Rays, 26, **99.**
Rays (of fins). See spines.
Regalecus jacksoniensis, 20.
Retropinna richardsoni, 22.
Rhina squatina, 25, 93, 98.
**Rhinobatidæ,** 11, 25.
**Rhinidæ,** 11, 25.
Rhinobatus, 13, 25, 97.
———— granulatus, 25, **97.**
Rhombosolea, 12, 77, 80.
Riverina fluviatilis, 13, 15, **104.**
Rock cod, 11, 14, **33.**
———— fish, 11.
Roe, function of, 8 ; position of, 8.

## S

Saccarius, 12.
Salmo, 35, 84, 155.
———— trutta, 84, 155.
———— salar, 9, 84, 155.
———— fario, 9, 155, 157,
———— salvelinus, 9.
\* Salmon of N.S.W., **35.**
**Salmonidæ,** 11, 22, 84.
Samson-fish, **58.**
Saurus, 22, 28,
———— australis, 22.
———— myops, 22.
———— nebulosa, 22.
———— truculenta, 22.
Saw-shark, 11, 93, **98.**
Scatophagus argus, 15.
———— multifasciatus, 15.
\* Schnapper, range of, 40.
———— described, 39.
———— fishing-grounds for, 41.
———— bait for, 42.
**Sciænidæ,** 11, 53.
Scientific descriptions, how mastered, 8.
Sciæna antarctica, 12, 17, **53.**
Sclerodermi, 11, 24, 90.
\* Scomber, 62.
———— antarcticus, 18, 62.
———— australasicus, 63.
———— pneumatophorus, 62.
**Scombresocidæ,** 11, 22, 83.
**Scrombridæ,** 11, 18, **57,** 61, 64.

Shark, 23, 91–97.
—— angel, 93, 98.
—— blind, 93, 97.
—— blue, 25, 92.
—— grey nurse, 91, 95.
—— hammer-head, 25, 92, 97.
—— sand, 97.
—— saw, 93, 98.
—— school, 25, 92.
—— shovel-nosed, 25, 97.
Scombresox, 81.
**Scopelidæ**, 11, 22, 82.
———————— teeth of the, 28.
**Scorpænidæ**, 11, 16, 47, 108.
Scorpæna cruenta, 13, 16, 47.
—————— bynœnsis, 16.
—————— cardinalis, 16.
* Scorpis æquipennis, 13, 15, 37.
* Scylla, 123.
Scyllium maculatum, 23.
**Scyllidæ**, 11, 23.
Sea-horse, 23, 28, 29.
* Sergeant Baker, 11, 22, 82.
Sebastes, 11.
—————— percoides, 16, 47.
Siphonognathus, 12.
Sepia, 122.
Sepioteuthis australis, 122.
* Seriola lalandi, 17, 59.
—————— nigrofasciatus, 17.
—————— grandis, 17.
—————— hippos, 17, 57, 60.
Serranus guttulatus, 12, 14.
—————— hexagonathus, 14.
—————— damellii, 14, 33.
—————— undulato-striatus, 14.
Sex in fishes, 8.
—— oysters, to distinguish, 112.
Shags, destruction of, 178.
Sillago, 12, 63.
* —————— ciliata, 65.
—————— bassensis, 18, 65.
* —————— maculata, 18, 65.
—————— punctata, 65.
**Siluridæ**, 5, 11, 13, 22, 80, 103.
—————— New South Wales, species of, 11, 22.
Silurus, 6.
Skeleton, value in classification, 5.
—————— of palæichthes, 6.
Skull of palæichthes, 6.
* Sole, 7, 22, 77.
* Solea, 21, 77.

South Australia, genera peculiar to, 12.
—————————— genera common to, and New South Wales, 12.
**Sparidæ**, 11, 16, 38.
Sphyræna, 12 ; Novæ-hollandiæ, 20, 69.
* —————— obtusata, 20, 69.
**Sphyrænidæ**, 11, 20, 69.
Spinacidæ, 12, 23.
Spines, homocanth, 6.
—————— hetrocanth, 6.
Spratelloides, 86.
Squamipinnes, 11, 37.
Squid, 122.
Star-fish, 122.
Sticharium, 20.
Stigmatophora, 12.
Stomach, forms of, 5.
Stranger, 75.
**Stromateidæ**, 62.
* Sub-operculum, 8, 9.
Sucking-fish, 63.
Sun-fish, 24.
* Sweep, 15, 37.
Sword-fish.
*Sydney Mail*, 2.
**Symbranchidæ**, 11, 23.
Synaptura, 12.
* —————— nigra, 23, 77.
Synacidium horridum, 16.
**Synganathidæ**, 11, 23.


T

Tail, heterocercal, 6.
—— homocercal, 6.
* Tarwhine, 16, 42.
Teleosti, 3, 14, 30.
—————— optic nerve of, 4.
* Temnodon saltator, 17, 57, 61.
Tenison-Woods, species discovered by, 12.
Tephræops, 12.
* Teraglin, 54.
Teratorhombus, 12.
Tetrodon, 12, 24, 90.
Teuthis javus, 12, 16, 51.
—————— nebulosa, 16, 50.
Therapon, 15, 104.
—————— cuverii, 15.
—————— niger, 15, 104.
—————— richardsoni, 15, 104.

Therapon unicolor, 15.
Thompson, Mr. F. A., 103.
Thoracici, 7.
Thornback, 26.
Thymallus australis, **109.**
Thymnus affinis, 18.
——— pelamys, 18, 64.
Thyrsites, 5, 13, **56.**
——— distribution of, 56.
Toad-fish, **90, 91.**
——— poisonous, 91.
**Torpedinidæ,** 11, 25.
Trachelochismus, 12.
Trachicthys, 12, 13.
Trachinidæ, 11, 18, 64.
Trachinops, 15.
Trachurus, 13, 58.
——— trachurus, 17, 57.
*——— declivis, **58.**
Trachynotus ovatus, 17.
——— bailloni, 17.
Trevally, the black, 16, **50.**
——— spines venomous, 49, 50.
——— the white, 17, 59.
**Trichiuridæ,** 11, 17, **56.**
Trichiurus haumela, 17.
Trigla, 12, 13, 38, 66, **68.**
——— kumu, 19, 66, **68.**
——— pleuracanthica, 19.
* ——— polyommata, 19, 68.
Trochocopus, 21.
Trochocochleæ, 122.
Trumpet-fish, 11.
Trumpeter, **47.**
Trygon, 13.
**Trygonidæ,** 11
Trygonorhina, 12, 25, **101.**
Tunisian fisheries, 123.

## V

Valenciennes, 2.
Vertebræ, notochord of, 6.
——— sometimes hard to distinguish, 5, 7.

## W

Weapons of attack, 7.
* Whiting, 7, 65.
——— rock, 21.
——— Melbourne, 65.
——— trumpeter, 18, 65.
* Wirrah, **34.**
Wobbegong, **25, 94.**
Wrasses, 7, **28.**

## X

Xiphiidæ, 11, 17, 55.

## Y

Yarra herring, **109.**
* Yellow-tail, 17, 57, **58.**
Young fishes mistaken for new species, 8.

## Z

* Zeus australis, 13, 18, **61.**
——— faber, 7, 11, **61.**
Zygæna, 12.
——— malleus, 22, 91, **97.**

---

Sydney : Thomas Richards, Government Printer.—1882.